Pitman Research Notes in Mathematics Series

Submission of proposals for consideration
Suggestions for publication, in the form of outlines and representative samples, are invited by the Editorial Board for assessment. Intending authors should approach one of the main editors or another member of the Editorial Board, citing the relevant AMS subject classifications. Alternatively, outlines may be sent directly to the publisher's offices. Refereeing is by members of the board and other mathematical authorities in the topic concerned, throughout the world.

Preparation of accepted manuscripts
On acceptance of a proposal, the publisher will supply full instructions for the preparation of manuscripts in a form suitable for direct photo-lithographic reproduction. Specially printed grid sheets can be provided and a contribution is offered by the publisher towards the cost of typing. Word processor output, subject to the publisher's approval, is also acceptable.

Illustrations should be prepared by the authors, ready for direct reproduction without further improvement. The use of hand-drawn symbols should be avoided wherever possible, in order to maintain maximum clarity of the text.

The publisher will be pleased to give any guidance necessary during the preparation of a typescript, and will be happy to answer any queries.

Important note
In order to avoid later retyping, intending authors are strongly urged not to begin final preparation of a typescript before receiving the publisher's guidelines. In this way it is hoped to preserve the uniform appearance of the series.

Addison Wesley Longman Ltd
Edinburgh Gate
Harlow, Essex, CM20 2JE
UK
(Telephone (0) 1279 623623)

Titles in this series. A full list is available from the publisher on request.

251 Stability of stochastic differential equations with respect to semimartingales
X Mao
252 Fixed point theory and applications
J Baillon and M Théra
253 Nonlinear hyperbolic equations and field theory
M K V Murthy and S Spagnolo
254 Ordinary and partial differential equations. Volume III
B D Sleeman and R J Jarvis
255 Harmonic maps into homogeneous spaces
M Black
256 Boundary value and initial value problems in complex analysis: studies in complex analysis and its applications to PDEs 1
R Kühnau and W Tutschke
257 Geometric function theory and applications of complex analysis in mechanics: studies in complex analysis and its applications to PDEs 2
R Kühnau and W Tutschke
258 The development of statistics: recent contributions from China
X R Chen, K T Fang and C C Yang
259 Multiplication of distributions and applications to partial differential equations
M Oberguggenberger
260 Numerical analysis 1991
D F Griffiths and G A Watson
261 Schur's algorithm and several applications
M Bakonyi and T Constantinescu
262 Partial differential equations with complex analysis
H Begehr and A Jeffrey
263 Partial differential equations with real analysis
H Begehr and A Jeffrey
264 Solvability and bifurcations of nonlinear equations
P Drábek
265 Orientational averaging in mechanics of solids
A Lagzdins, V Tamuzs, G Teters and A Kregers
266 Progress in partial differential equations: elliptic and parabolic problems
C Bandle, J Bemelmans, M Chipot, M Grüter and J Saint Jean Paulin
267 Progress in partial differential equations: calculus of variations, applications
C Bandle, J Bemelmans, M Chipot, M Grüter and J Saint Jean Paulin
268 Stochastic partial differential equations and applications
G Da Prato and L Tubaro
269 Partial differential equations and related subjects
M Miranda
270 Operator algebras and topology
W B Arveson, A S Mishchenko, M Putinar, M A Rieffel and S Stratila
271 Operator algebras and operator theory
W B Arveson, A S Mishchenko, M Putinar, M A Rieffel and S Stratila
272 Ordinary and delay differential equations
J Wiener and J K Hale
273 Partial differential equations
J Wiener and J K Hale
274 Mathematical topics in fluid mechanics
J F Rodrigues and A Sequeira
275 Green functions for second order parabolic integro-differential problems
M G Garroni and J F Menaldi
276 Riemann waves and their applications
M W Kalinowski
277 Banach C(K)-modules and operators preserving disjointness
Y A Abramovich, E L Arenson and A K Kitover
278 Limit algebras: an introduction to subalgebras of C*-algebras
S C Power
279 Abstract evolution equations, periodic problems and applications
D Daners and P Koch Medina
280 Emerging applications in free boundary problems
J Chadam and H Rasmussen
281 Free boundary problems involving solids
J Chadam and H Rasmussen
282 Free boundary problems in fluid flow with applications
J Chadam and H Rasmussen
283 Asymptotic problems in probability theory: stochastic models and diffusions on fractals
K D Elworthy and N Ikeda
284 Asymptotic problems in probability theory: Wiener functionals and asymptotics
K D Elworthy and N Ikeda
285 Dynamical systems
R Bamon, R Labarca, J Lewowicz and J Palis
286 Models of hysteresis
A Visintin
287 Moments in probability and approximation theory
G A Anastassiou
288 Mathematical aspects of penetrative convection
B Straughan
289 Ordinary and partial differential equations. Volume IV
B D Sleeman and R J Jarvis
290 *K*-theory for real *C**-algebras
H Schröder
291 Recent developments in theoretical fluid mechanics
G P Galdi and J Necas
292 Propagation of a curved shock and nonlinear ray theory
P Prasad
293 Non-classical elastic solids
M Ciarletta and D Ieșan
294 Multigrid methods
J Bramble
295 Entropy and partial differential equations
W A Day
296 Progress in partial differential equations: the Metz surveys 2
M Chipot
297 Nonstandard methods in the calculus of variations
C Tuckey
298 Barrelledness, Baire-like- and (LF)-spaces
M Kunzinger
299 Nonlinear partial differential equations and their applications. Collège de France Seminar. Volume XI
H Brezis and J L Lions

300 Introduction to operator theory
T Yoshino
301 Generalized fractional calculus and applications
V Kiryakova
302 Nonlinear partial differential equations and their applications. Collège de France Seminar Volume XII
H Brezis and J L Lions
303 Numerical analysis 1993
D F Griffiths and G A Watson
304 Topics in abstract differential equations
S Zaidman
305 Complex analysis and its applications
C C Yang, G C Wen, K Y Li and Y M Chiang
306 Computational methods for fluid-structure interaction
J M Crolet and R Ohayon
307 Random geometrically graph directed self-similar multifractals
L Olsen
308 Progress in theoretical and computational fluid mechanics
G P Galdi, J Málek and J Necas
309 Variational methods in Lorentzian geometry
A Masiello
310 Stochastic analysis on infinite dimensional spaces
H Kunita and H-H Kuo
311 Representations of Lie groups and quantum groups
V Baldoni and M Picardello
312 Common zeros of polynomials in several variables and higher dimensional quadrature
Y Xu
313 Extending modules
N V Dung, D van Huynh, P F Smith and R Wisbauer
314 Progress in partial differential equations: the Metz surveys 3
M Chipot, J Saint Jean Paulin and I Shafrir
315 Refined large deviation limit theorems
V Vinogradov
316 Topological vector spaces, algebras and related areas
A Lau and I Tweddle
317 Integral methods in science and engineering
C Constanda
318 A method for computing unsteady flows in porous media
R Raghavan and E Ozkan
319 Asymptotic theories for plates and shells
R P Gilbert and K Hackl
320 Nonlinear variational problems and partial differential equations
A Marino and M K V Murthy
321 Topics in abstract differential equations II
S Zaidman
322 Diffraction by wedges
B Budaev
323 Free boundary problems: theory and applications
J I Diaz, M A Herrero, A Liñan and J L Vazquez
324 Recent developments in evolution equations
A C McBride and G F Roach
325 Elliptic and parabolic problems: Pont-à-Mousson 1994
C Bandle, J Bemelmans, M Chipot, J Saint Jean Paulin and I Shafrir
326 Calculus of variations, applications and computations: Pont-à-Mousson 1994
C Bandle, J Bemelmans, M Chipot, J Saint Jean Paulin and I Shafrir
327 Conjugate gradient type methods for ill-posed problems
M Hanke
328 A survey of preconditioned iterative methods
A M Bruaset
329 A generalized Taylor's formula for functions of several variables and certain of its applications
J-A Riestra
330 Semigroups of operators and spectral theory
S Kantorovitz
331 Boundary-field equation methods for a class of nonlinear problems
G N Gatica and G C Hsiao
332 Metrizable barrelled spaces
J C Ferrando, M López Pellicer and L M Sánchez Ruiz
333 Real and complex singularities
W L Marar
334 Hyperbolic sets, shadowing and persistence for noninvertible mappings in Banach spaces
B Lani-Wayda
335 Nonlinear dynamics and pattern formation in the natural environment
A Doelman and A van Harten
336 Developments in nonstandard mathematics
N J Cutland, V Neves, F Oliveira and J Sousa-Pinto
337 Topological circle planes and topological quadrangles
A E Schroth
338 Graph dynamics
E Prisner
339 Localization and sheaves: a relative point of view
P Jara, A Verschoren and C Vidal
340 Mathematical problems in semiconductor physics
P Marcati, P A Markowich and R Natalini
341 Surveying a dynamical system: a study of the Gray–Scott reaction in a two-phase reactor
K Alhumaizi and R Aris
342 Solution sets of differential equations in abstract spaces
R Dragoni, J W Macki, P Nistri and P Zecca
343 Nonlinear partial differential equations
A Benkirane and J-P Gossez
344 Numerical analysis 1995
D F Griffiths and G A Watson
345 Progress in partial differential equations: the Metz surveys 4
M Chipot and I Shafrir
346 Rings and radicals
B J Gardner, Liu Shaoxue and R Wiegandt
347 Complex analysis, harmonic analysis and applications
R Deville, J Esterle, V Petkov, A Sebbar and A Yger
348 The theory of quantaloids
K I Rosenthal
349 General theory of partial differential equations and microlocal analysis
Qi Min-you and L Rodino

350 Progress in elliptic and parabolic partial differential equations
A Alvino, P Buonocore, V Ferone, E Giarrusso, S Matarasso, R Toscano and G Trombetti

351 Integral representations for spatial models of mathematical physics
V V Kravchenko and M V Shapiro

352 Dynamics of nonlinear waves in dissipative systems: reduction, bifurcation and stability
G Dangelmayr, B Fiedler, K Kirchgässner and A Mielke

353 Singularities of solutions of second order quasilinear equations
L Véron

354 Mathematical theory in fluid mechanics
G P Galdi, J Málek and J Necas

G P Galdi
University of Ferrara, Italy
J Málek
Charles University, Czech Republic
and
J Nečas
Charles University, Czech Republic
(Editors)

Mathematical theory in fluid mechanics

LONGMAN

Addison Wesley Longman Limited
Edinburgh Gate, Harlow
Essex CM20 2JE, England
and Associated Companies throughout the world.

Published in the United States of America
by Addison Wesley Longman Inc.

First published 1996

AMS Subject Classifications: (Main) 35Q, 76A-D
(Subsidiary) 76E

ISSN 0269-3674

ISBN 0 582 29810 5

British Library Cataloguing in Publication Data

A catalogue record for this book is
available from the British Library

Printed and bound in Great Britain
by Biddles Ltd, Guildford and King's Lynn

Contents

Foreword

This volume consists of four contributions that are based on a series of lectures delivered by Jens Frehse, Konstantin Pileckas, K.R. Rajagopal and Wolf von Wahl at the Fourth Winter School in Mathematical Theory in Fluid Mechanics, held at Paseky, Czech Republic, in December 3 – December 9, 1995. In these papers the authors present recent contributions and updated surveys of relevant topics in the various areas of theoretical fluid mechanics.

The first contribution written by Jens Frehse and Michael Růžička is devoted to the question of the existence of a regular solution to stationary Navier-Stokes equations in five dimensions. Some of applied techniques are also used to construct a solution to stationary Navier-Stokes equations in $\mathbb{R}^3$ with fast decay at infinity. The paper of Konstantin Pileckas surveys recent results regarding the solvability of the Stokes and Navier-Stokes system in domains with outlets at infinity. The third contribution presented by K.R. Rajagopal introduces a continuum approach to mixture theory with the emphasis to the constitutive equations, boundary conditions and moving singular surface. The last paper of Ralf Kaiser and Wolf von Wahl brings new results on stability of basic flow for the Taylor-Couette problem in the small-gap limit.

It is our belief that this volume will furnish new insights into classical problems as well as significant ideas on current topics.

We would like to thank all participants for their interest and for their stimulating questions and discussions during and after the lectures. We would also like to acknowledge unselfish help of the other organizers, Antonín Novotný, Mirko Rokyta and Michael Růžička. Finally we thank Gudrun Thäter, Michaela Lichá and Jaroslav Hron for their help to prepare the camera–ready version.

May 1996

Giovanni P. Galdi
Josef Málek
Jindřich Nečas

JENS FREHSE and MICHAEL RŮŽIČKA

Weighted Estimates for the Stationary Navier-Stokes Equations

1. Introduction

In this paper we present several methods to obtain local weighted estimates for solutions of the stationary Navier-Stokes equations

$$\begin{aligned} -\Delta \mathbf{u} + \mathbf{u} \cdot \nabla \mathbf{u} + \nabla p &= \mathbf{f} && \text{in } \Omega \\ \operatorname{div} \mathbf{u} &= 0 && \\ \mathbf{u} &= \mathbf{0} && \text{on } \partial\Omega \end{aligned} \tag{1.1}$$

in a bounded domain Ω of $\mathbb{R}^n$, where $\mathbf{f} \in L^\infty(\Omega)$. By *weighted* estimates we mean estimates like

$$\int_{\Omega_0} |\mathbf{u}|^q |x - x_0|^{n-\lambda}\, dx \le K\,, \qquad q \ge 2 \tag{1.2}$$

uniformly with respect to $x_0 \in \Omega_0 \subset\subset \Omega$, and corresponding ones for the first and second derivatives of $\mathbf{u}$ and for p and its first derivatives.

These estimates were developed by the authors in several papers in order to contribute to the question of regularity of the stationary Navier-Stokes equations for dimension $n \ge 5$, furthermore we believe that these methods can be useful for several other problems, as already illustrated in the last section of this paper.

The study of the five dimensional case is motivated by the fact that the non-stationary Navier-Stokes equations in three space dimensions, where the problem of regularity of solutions is a famous unsolved problem, has certain similarities to the five dimensional stationary case. Among other things there are the following analogies for $H^1(\Omega)$ resp. $L^\infty(0,T;L^2(\Omega)) \cap L^2(0,T;H^1(\Omega))$ solutions of the stationary resp. non-stationary Navier-Stokes equations (cf. [21], [23], [30], [3], [20], [29], [8], [5], [2], [34])

stationary case $n=5$	nonstationary case $n=3$
$\mathbf{u} \in L^{10/3}(\Omega)$	$\mathbf{u} \in L^{10/3}(0,T;L^{10/3}(\Omega))$
$p \in L^{5/3}(\Omega)$	$p \in L^{5/3}(0,T;L^{5/3}(\Omega))$
$\nabla p \in L^{5/4}(\Omega)$	$\nabla p \in L^{5/4}(0,T;L^{5/4}(\Omega))$

$\nabla^2\mathbf{u} \in L^{4/3-\delta}(\Omega)$	$\nabla^2\mathbf{u} \in L^{4/3-\delta}(0,T;L^{4/3-\delta}(\Omega))$
$\mathbf{u} \in L^5(\Omega) \Rightarrow$ full regularity	$\mathbf{u} \in L^5(0,T;L^5(\Omega)) \Rightarrow$ full regularity
partial regularity, i.e. $\mathbf{u}$ is regular in a neighbourhood of any x_0 if $R^{-1}\int_{B_R(x_0)} \lvert\nabla\mathbf{u}\rvert^2\,dx$ is small	partial regularity, i.e. $\mathbf{u}$ is regular in a neighbourhood of any x_0 if $R^{-1}\iint_{Q_R(x_0)} \lvert\nabla\mathbf{u}\rvert^2\,dx\,dt$ is small.

We remark that for the step $\{\mathbf{u} \in L^5(0,T;L^5(\Omega)) \Rightarrow$ full regularity $\}$ there is an elegant proof of Sohr [29], von Wahl [36] and that in the stationary case the conclusion $\{\mathbf{u} \in L^q(\Omega) \Rightarrow$ full regularity $\}$ is known for q such that $q \geq 4$ and $q > \frac{n}{2}$, cf. [14].

Since the problem of local regularity of solutions to the stationary equations can be simply solved in dimension 2, 3 and 4 via Sobolev's imbedding theorem and the linear theory of the Stokes equation (cf. [24]) (a tiny additional borderline argument in the case $n = 4$, cf. [17]) there was some interest to study the stationary case for $n = 5$ and $n > 5$.

The first step concerning better regularity results and the basis for further progress was done in [9], where it was proved that for $n = 5$ there is a solution to the stationary problem such that for all $\delta > 0$, small,

$$\int_{\Omega_0} \lvert\mathbf{u}\rvert^2 \lvert x - x_0\rvert^{-3+\delta}\,dx \leq K\,, \tag{1.3}$$

uniformly with respect to $x_0 \in \Omega_0 \subset\subset \Omega$ and via a bootstrap argument, for some $\delta' > 0$,

$$\int_{\Omega_0} \lvert\mathbf{u}\rvert^{4-\delta} \lvert x - x_0\rvert^{-1+\delta'}\,dx \leq K\,, \tag{1.4}$$

thus for the first time one exceeded the mere conclusion coming from Sobolev's imbedding theorem.

Furthermore, a bootstrap argument in weighted spaces and the linear theory in weighted spaces (cf. Section 7 of these notes) allowed the authors to perform the step:

If

$$\int_{\Omega_0} \lvert\nabla\mathbf{u}\rvert^2 \lvert x - x_0\rvert^{-1-\delta}\,dx \leq K\,, \tag{1.5}$$

uniformly with respect to $x_0 \in \Omega_0 \subset\subset \Omega$,

then

$$\mathbf{u} \text{ is regular, say } \mathbf{u} \in H^{2,p}_{\text{loc}}(\Omega)\ \forall p < \infty\,.$$

Condition (1.5) is slightly stronger than

$$\int_{\Omega_0} \lvert\mathbf{u}\rvert^2 \lvert x - x_0\rvert^{-3-\delta}\,dx \leq K\,, \tag{1.6}$$

which can be shown to be also sufficient for the above implication. The previous implications work analogously in arbitrary dimensions $n \geq 5$:

If

$$\int_{\Omega_0} |\nabla \mathbf{u}|^2 |x - x_0|^{-n+4-\delta}\, dx \leq K\,, \tag{1.7}$$

uniformly with respect to $x_0 \in \Omega_0 \subset\subset \Omega$,

then

$$\mathbf{u} \text{ is regular}\,.$$

Also the condition analogous to (1.6) in dimension $n \geq 5$ is sufficient for the regularity. The proof uses two properties which are specific for the Navier-Stokes equations.

Let first $\mathbf{u}, p$ be smooth solutions of (1.1). Then we observe that there is a differential equation, a differential inequality and a maximum principle for the quantity

$$\frac{\mathbf{u}^2}{2} + p\,, \tag{1.8}$$

whose physical name is "klidový tlak" ("head pressure", "Ruhedruck"). The differential equation reads

$$-\Delta\Big(\frac{\mathbf{u}^2}{2} + p\Big) + \mathbf{u} \cdot \nabla\Big(\frac{\mathbf{u}^2}{2} + p\Big) = (\nabla \mathbf{u}) \cdot (\nabla \mathbf{u})^T - |\nabla \mathbf{u}|^2 + \mathbf{f} \cdot \mathbf{u} - \operatorname{div} \mathbf{f}\,,$$

hence ($|\nabla \mathbf{u}|^2 - (\nabla \mathbf{u}) \cdot (\nabla \mathbf{u})^T = |\operatorname{curl} \mathbf{u}|^2 \geq 0$)

$$-\Delta\Big(\frac{\mathbf{u}^2}{2} + p\Big) + \mathbf{u} \cdot \nabla\Big(\frac{\mathbf{u}^2}{2} + p\Big) \leq \mathbf{f} \cdot \mathbf{u} - \operatorname{div} \mathbf{f}\,. \tag{1.9}$$

From (1.9) one expects a bound

$$\frac{\mathbf{u}^2}{2} + p \leq K \tag{1.10}$$

on interior domains of Ω. Using (1.10) and merely the pressure equation (obtained by taking the divergence of (1.1)).

$$-\Delta p = D_i u_k D_k u_i - \operatorname{div} \mathbf{f}$$

the weighted estimates

$$\int_{\Omega_0} \Big|\frac{\mathbf{u}^2}{2} + p\Big| |x - x_0|^{2-n}\, dx \leq K\,,$$

$$\int_{\Omega_0} \mathbf{u}^2 |x - x_0|^{2-n+\delta}\, dx \leq K\,,$$

can be derived in any dimension. This is shown in Section 2 of the these notes.

Unfortunately, the quantity $u_i D_i(\frac{\mathbf{u}^2}{2} + p)$ appearing in (1.9) is not even in L^1, a priori. Thus there remains the question, whether there is a rigorous proof for (1.10). This was done in our first paper [9] by constructing approximate solutions using approximations of the Navier-Stokes equations with the additional term $\varepsilon \mathbf{u}|\mathbf{u}|^2$ and then passing to the limit $\varepsilon \to 0$, cf. Sections 3 and 4 of this paper. However, we succeeded to prove (1.10) rigorously only if $n = 5, 6$ for the Dirichlet problem, and if $n \leq 15$ for the periodic case, (cf. [8]–[14], [28]).

Having done this, there arose the question whether one can fill the "δ-gap" between (1.3) and (1.5). This was attacked in parallel papers by M. Struwe [35] who treated the problem in whole $\mathbb{R}^5$, and by the authors (cf. [10]) who dealt with the periodic case. Both papers use, for our taste, a comparatively delicate degree and deformation argument. The authors succeeded in [11] (cf. also Section 5) to replace the subtle arguments of [35] and [10] by an additional estimate which works in higher dimensions and covers also the (more difficult) case of a bounded domain.

In fact the authors were able to show that

<u>A local inequality</u>

$$\frac{\mathbf{u}^2}{2} + p \leq K$$

<u>implies full regularity for any dimension</u>.

This was proved by the so called hole filling technique which was used successfully in many other branches of elliptic analysis and is explained, for $n = 5$, in Section 6 of the these notes.

In the last part, Section 8, of these notes we apply the preceding techniques, to the study the decay properties of the Navier-Stokes equations in the whole of $\mathbb{R}^3$. We construct a solution for which

$$\int_{\mathbb{R}^3} |\mathbf{u}|^2 |x|^{-1-\delta}\, dx \leq K$$

and consequently

$$\int_{\mathbb{R}^3} |\nabla \mathbf{u}|^2 |x|^{1-\delta}\, dx \leq K\,.$$

These results are new and not published elsewhere, but use similar ideas as in the treatment of the problem of regularity of solutions in dimension $n \geq 5$, combined with additional new techniques.

Many parts of our proofs work also in the <u>instationary</u> case, (note that Section 2 holds also for the instationary case since one works merely with the pressure equation), however we unfortunately did not find up to now an analogous quantity for "klidový tlak" in the nonstationary case.

2. First weighted estimates derived from an upper bound for the head pressure

Let us start with simplest part of the theory. It consists in the conclusion, that an estimate for the positive part of the so called "klidový tlak" (head pressure)

$$\frac{\mathbf{u}^2}{2} + p$$

implies a weighted estimate for $\mathbf{u}^2$ and $|p|$, provided that p satisfies the pressure inequality

$$-\Delta p \geq D_i(u_k D_k u_i) - \operatorname{div} \mathbf{f} \tag{2.1}$$

in the weak sense and that $\operatorname{div} \mathbf{u} = 0$. Note that no further equation besides the pressure inequality is required to hold, in particular the Navier-Stokes equations are not required to hold in the following proposition.

Proposition 2.2. *Let $p \in L^1(\Omega)$, $\mathbf{u} \in H^1(\Omega)$ satisfy the pressure inequality (2.1) and let $\operatorname{div} \mathbf{u} = 0$ and $\int_\Omega \frac{|\mathbf{f}|}{|x-x_0|^{s-3}} dx < \infty$. Suppose that $4 < s \leq n$ and*

$$\int_{\Omega_0} \left(\frac{\mathbf{u}^2}{2} + p\right)_+ \frac{1}{|x-x_0|^{s-2}}\, dx < \infty\,.$$

Then there is a constant K_n depending only on n such that

$$\begin{aligned}
&\int_{\Omega_0} \left|\frac{\mathbf{u}^2}{2} + p\right| \frac{1}{|x-x_0|^{s-2}}\, dx + (n-s) \int_{\Omega_0} \mathbf{u}^2 \frac{1}{|x-x_0|^{s-2}}\, dx \\
&\leq K_n \int_\Omega \left(\frac{\mathbf{u}^2}{2} + p\right)_+ \frac{1}{|x-x_0|^{s-2}}\, dx + K_n \int_\Omega |\mathbf{f}||x-x_0|^{3-s}\, dx \\
&\quad + K_n |\operatorname{dist}(x_0, \partial\Omega)|^{-s} \left(\|p\|_{L^1(\Omega)} + \|\mathbf{u}\|^2_{L^2(\Omega)}\right) .
\end{aligned}$$

Remark 2.3. Clearly the assumption on $\mathbf{f}$ is satisfied if $\mathbf{f} \in L^{\frac{n}{n-3}-\delta}(\Omega)$. So, if $\frac{\mathbf{u}^2}{2} + p \leq K$ we have $\int_{\Omega_0} \left(\frac{\mathbf{u}^2}{2} + p\right)_+ \frac{1}{|x-x_0|^{n-2}}\, dx \leq K$ and Proposition 2.2 gives us an estimate for $\int_{\Omega_0} \left|\frac{\mathbf{u}^2}{2} + p\right| \frac{1}{|x-x_0|^{n-2}}\, dx$ and $\int_{\Omega_0} \mathbf{u}^2 \frac{1}{|x-x_0|^{n-2-\varepsilon}}\, dx, \varepsilon > 0$ if, say $\mathbf{f} \in L^{\frac{n}{n-3}-\delta}(\Omega)$.

Proof of Proposition 2.2: Let $\zeta \in C_0^\infty(\Omega)$ be a truncation function, $\zeta = 1$ on $\Omega_1 \subset\subset \Omega$, $\Omega_0 \subset\subset \Omega_1$. For inequality (2.1) we choose the function

$$\zeta(x) \frac{1}{(|x-x_0|^2 + h^2)^{(s-4)/2}}$$

as a test function and obtain by means of integration by parts

$$-\int_\Omega p\Delta\left(\zeta\frac{1}{(|x-x_0|^2+h^2)^{(s-4)/2}}\right)dx \geq \int_\Omega u_iu_kD_iD_k\left(\zeta\frac{1}{(|x-x_0|^2+h^2)^{(s-4)/2}}\right)dx$$
$$+\int_\Omega \mathbf{f}\cdot\nabla\left(\zeta\frac{1}{(|x-x_0|^2+h^2)^{(s-4)/2}}\right)dx\,.$$

Performing the differentiations by Leibniz's rule, there arise terms with the factor $D_j\zeta$, $D_iD_k\zeta$ which vanish on Ω_1 since $\zeta = 1$ on Ω_1. Since $x_0 \in \Omega_0 \subset\subset \Omega_1$ these terms contribute terms like p and u_iu_k multiplied by a function which is uniformly bounded in $L^\infty(\Omega)$ as $h \to 0$ and x_0 varies in Ω_0. These pollution terms can be estimated by the term $K_n\,\mathrm{dist}(\partial\Omega, x_0)^{-s}\left(\|p\|_{L^1(\Omega)} + \|u\|^2_{L^2(\Omega)}\right)$. The term $\int_\Omega \mathbf{f}\,\nabla\left(\zeta(|x-x_0|^2+h^2)^{(4-s)/2}\right)dx$ is uniformly bounded for $h \to 0$, $x_0 \in \Omega_0$ if $\mathbf{f} \in L^{\frac{n}{n-3}-\delta}(\Omega)$, $s \leq n$, since $\nabla\frac{1}{|x-x_0|^{s-4}} \in L^{\frac{n}{3}+\delta}(\Omega)$ for $s = n$. Thus

$$\int_\Omega \zeta\left\{p\Delta\frac{1}{(|x-x_0|^2+h^2)^{(s-4)/2}} + u_iu_kD_iD_k\frac{1}{(|x-x_0|^2+h^2)^{(s-4)/2}}\right\}dx$$
$$\leq K\int_{\Omega_1}\left\{|\mathbf{u}|^2+|p|\right\}dx + K$$

uniformly with respect to $h \to 0$, $x_0 \in \Omega_0$. By a simple calculation

$$\Delta\frac{1}{(|x-x_0|^2+h^2)^{(s-4)/2}} = -(s-4)\frac{(n-s+2)|x-x_0|^2+nh^2}{(|x-x_0|^2+h^2)^{s/2}}$$
$$u_iu_kD_iD_k\frac{1}{(|x-x_0|^2+h^2)^{(s-4)/2}} = -(s-4)\frac{\mathbf{u}^2}{(|x-x_0|^2+h^2)^{s-2/2}}$$
$$+(s-4)(s-2)\frac{|\mathbf{u}\cdot(x-x_0)|^2}{(|x-x_0|^2+h^2)^{s/2}}$$
$$= -(s-4)\frac{|x-x_0|^2+h^2}{(|x-x_0|^2+h^2)^{s/2}}\mathbf{u}^2$$
$$+(s-4)(s-2)\frac{|\mathbf{u}\cdot(x-x_0)|^2}{(|x-x_0|^2+h^2)^{s/2}}\,.$$

Hence we obtain

$$(s-4)\int_{\Omega_0}\frac{|x-x_0|^2}{(|x-x_0|^2+h^2)^{s/2}}\left\{-(\frac{\mathbf{u}^2}{2}+p)(n-s+2)+\mathbf{u}^2(\frac{n-s+2}{2}-1)\right\}dx$$
$$+(s-4)\int_{\Omega_0}\frac{h^2}{(|x-x_0|^2+h^2)^{s/2}}\left\{-(\frac{\mathbf{u}^2}{2}+p)n+\mathbf{u}^2(\frac{n}{2}-1)\right\}dx$$
$$+(s-4)(s-2)\int_{\Omega_0}\frac{1}{(|x-x_0|^2+h^2)^{s/2}}|\mathbf{u}\cdot(x-x_0)|^2\,dx \leq K\,.$$

We use that $-(p+\frac{\mathbf{u}^2}{2}) = |p+\frac{\mathbf{u}^2}{2}| - 2\big(p+\frac{\mathbf{u}^2}{2}\big)_+$ and drop nonnegative summands arising from the second integral. This yields, canceling the factor $s-4$,

$$\int_{\Omega_0} \frac{|x-x_0|^2}{(|x-x_0|^2+h^2)^{s/2}} \left\{ |\frac{\mathbf{u}^2}{2}+p|(n-s+2) + \mathbf{u}^2(\frac{n-s+2}{2}-1) \right\} dx$$
$$+ (s-2)\int_{\Omega_0} \frac{1}{(|x-x_0|^2+h^2)^{s/2}} |\mathbf{u}\cdot(x-x_0)|^2\, dx$$
$$\le K + 2(n-s+2)\int_{\Omega_0} \frac{|x-x_0^2| + K_0 h^2}{(|x-x_0|^2+h^2)^{s/2}} \big(\frac{\mathbf{u}^2}{2}+p\big)_+\, dx\,,$$

where $K_0 = n/(n-s+2)$. For $s \le n$, the factor $\frac{n-s+2}{2} - 1 = \frac{1}{2}(n-s) \ge 0$. Furthermore

$$\frac{|x-x_0|^2+K_0h^2}{(|x-x_0|^2+h^2)^{s/2}} \le K_0 \frac{1}{(|x-x_0|^2+h^2)^{(s-2)/2}}\,.$$

Hence we may pass to the limit $h \to 0$ and obtain

$$\int_{\Omega_0} |\frac{\mathbf{u}^2}{2}+p| \frac{n-s+2}{|x-x_0|^{s-2}} + \frac{\mathbf{u}^2}{2}\frac{n-s}{|x-x_0|^{s-2}}\, dx + (s-2)\int_{\Omega_0} \frac{|\mathbf{u}\cdot(x-x_0)|^2}{|x-x_0|^s}\, dx$$
$$\le K + K_n \int_{\Omega} \frac{\big(\frac{\mathbf{u}^2}{2}+p\big)_+}{|x-x_0|^{s-2}}\, dx\,.$$

This proves Proposition 2.2. ■

3. Approximate solutions

As we have remarked in the introductory section, the differential inequality (1.9) for the "klidový tlak" $\frac{\mathbf{u}^2}{2}+p$ seems to imply a bound for this quantity from above as needed in Proposition 2.2 (cf. Remark 2.3). However one has a problem due to the lack of regularity of $\mathbf{u}$ unless one assumes additional regularity (eg. $\mathbf{u}\in L^4(\Omega), n \le 15$ (periodic case) see [14]), which we do not want to do here. Thus one confines to construct a solution such that $\frac{\mathbf{u}^2}{2}+p \le K$. This is done in this and the following section. Since one wants to work with the weak form of the head pressure inequality one needs that $u_iD_i\big(\frac{\mathbf{u}^2}{2}+p\big) \in L^1(\Omega)$ which is true if $\mathbf{u}\in L^4(\Omega), p\in L^2(\Omega)$ and $\nabla\mathbf{u} \in L^2(\Omega)$. This motivates why we choose an approximation whose solutions are in $L^4(\Omega)$. Approximate problem: Find $\mathbf{u}_\varepsilon \in H_0^1(\Omega)\cap L^4(\Omega)$ such that for all $\boldsymbol{\varphi}\in H_0^1(\Omega)\cap L^4(\Omega), \operatorname{div}\boldsymbol{\varphi} = 0$

$$\big(\nabla\mathbf{u}^\varepsilon, \nabla\boldsymbol{\varphi}\big) + \big(\mathbf{u}^\varepsilon\cdot\nabla\mathbf{u}^\varepsilon, \boldsymbol{\varphi}\big) + \varepsilon\big(|\mathbf{u}^\varepsilon|^2\mathbf{u}^\varepsilon, \boldsymbol{\varphi}\big) = \big(\mathbf{f}, \boldsymbol{\varphi}\big)$$
$$\operatorname{div}\mathbf{u}^\varepsilon = 0\,.$$

It is clear that such a solution exists, say if $\mathbf{f}\in L^{\frac{2n}{n+2}}(\Omega)$ this is done via a Ritz-Galerkin approach using the coercivity of the form in $H_0^1(\Omega)\cap L^4(\Omega)\cap\{\mathbf{v}|\operatorname{div}\mathbf{v}=0\}$. It is a simple exercise to prove the following proposition.

Proposition 3.1. *For $\varepsilon \to 0$, there exists a subnet $\{\mathbf{u}^\varepsilon\}_{\varepsilon\in\Lambda}$ such that $\mathbf{u}^\varepsilon \rightharpoonup \mathbf{u}$ weakly in $H_0^1(\Omega)$ and $\mathbf{u}$ is a weak solution of the Navier-Stokes equations (1.1). Furthermore $\nabla p^\varepsilon \rightharpoonup \nabla p$ weakly in $L^{n/(n-1)}_{\text{loc}}(\Omega)$, p being the corresponding pressure.*

It turns out that also a one sided bound for our key-quantity "klidový tlak" can be established. The remainder of this section is devoted to this problem.

The approximate pressure equation reads

$$-\Delta p^\varepsilon = D_i u_k^\varepsilon D_k u_i^\varepsilon + \varepsilon \operatorname{div} |\mathbf{u}^\varepsilon|^2 \mathbf{u}^\varepsilon - \operatorname{div} \mathbf{f}\,,$$

and the head pressure equation reads

$$\begin{aligned}-\Delta\Big(\frac{|\mathbf{u}^\varepsilon|^2}{2} + p^\varepsilon\Big) + \mathbf{u}^\varepsilon \cdot \nabla\Big(\frac{|\mathbf{u}^\varepsilon|^2}{2} + p^\varepsilon\Big) + |\mathbf{u}^\varepsilon|^4 + |\nabla \mathbf{u}^\varepsilon|^2 - (\nabla u^\varepsilon)\cdot(\nabla u^\varepsilon)^T \\ = \mathbf{f}\cdot\mathbf{u}^\varepsilon + \varepsilon \operatorname{div}|\mathbf{u}^\varepsilon|^2\mathbf{u}^\varepsilon - \operatorname{div}\mathbf{f}\,.\end{aligned} \tag{3.2}$$

This holds in the weak sense, the space $L^\infty(\Omega) \cap H_0^1(\Omega)$ with compact support is an admissible class of test functions since $|\mathbf{u}^\varepsilon|^2 \nabla \mathbf{u}^\varepsilon \in L^1(\Omega)$, $\nabla p^\varepsilon \in L^{4/3}_{\text{loc}}(\Omega)$ (note $D_i u_k^\varepsilon D_k u_i^\varepsilon = D_i(u_k^\varepsilon D_k u_i^\varepsilon)$, $\mathbf{u}^\varepsilon \nabla \mathbf{u}^\varepsilon \in L^{4/3}(\Omega)$). In the next section, we shall prove the existence of an auxiliary function $G_h \in H_0^1(\Omega) \cap L^\infty(\Omega)$, $G_h \geq 0$ such that the "dual equation"

$$-\Delta G_h - u_i^\varepsilon D_i G_h = \delta_h \tag{3.3}$$

holds weakly in Ω, $G_h = G_{h,\varepsilon}$, where δ_h will mostly be defined as $\delta_h = |B_h|^{-1}$ on $B_h(x_0)$, however in this moment, $\delta_h \geq 0$ and $\delta_h \in L^\infty(\Omega)$ is sufficient. The test functions range in $H_0^1(\Omega) \cap L^{\frac{4}{3}}(\Omega)$ if the equation is written in the weak sense

$$\int_\Omega \nabla G_h \nabla \varphi\, dx + \int_\Omega \nabla G_h \cdot \mathbf{u}^\varepsilon \varphi\, dx = \int_\Omega \delta_h\, \varphi\, dx\,.$$

In (3.2), we use the mollification $\omega_r * (\tau G_h)$, τ being a nonnegative localization function, and obtain, using the properties of $G_h (\geq 0)$,

$$\begin{aligned}&-\int_\Omega \omega_r * \Big(\frac{|\mathbf{u}^\varepsilon|^2}{2} + p^\varepsilon\Big)\tau \Delta G_h\, dx \\ &\quad + \int_\Omega \omega_r * \Big[\mathbf{u}^\varepsilon \cdot \nabla\Big(\frac{|\mathbf{u}^\varepsilon|^2}{2} + p^\varepsilon\Big)\Big]\tau G_h\, dx + \varepsilon \int_\Omega \omega_r * |\mathbf{u}^\varepsilon|^4 \tau G_h\, dx \\ &\leq \int_\Omega \big|\mathbf{f}\cdot\mathbf{u}^\varepsilon + \varepsilon \operatorname{div}|\mathbf{u}^\varepsilon|^2\mathbf{u}^\varepsilon - \operatorname{div}\mathbf{f}\big|\, \omega_r * \tau G_h\, dx \\ &\quad + \text{terms containing } \nabla\tau \text{ and } \nabla^2\tau\,,\end{aligned} \tag{3.4}$$

which have better L^q-properties since τ may be chosen smooth.

We are allowed to pass to the limit $r \to 0$ of the convolution parameter except in the first term of (3.4) which is written as (note $\nabla^2 G_h \in L^1(\Omega)$)

$$-\int_\Omega \omega_r * (\frac{|\mathbf{u}^\varepsilon|^2}{2} + p^\varepsilon)\tau\Delta G_h\,dx = \int_\Omega \omega_r * (\frac{|\mathbf{u}^\varepsilon|^2}{2} + p^\varepsilon)\tau\delta_h\,dx$$
$$-\int_\Omega \mathbf{u}^\varepsilon \cdot \nabla(\omega_r * [\frac{|\mathbf{u}^\varepsilon|^2}{2} + p^\varepsilon]\tau)G_h\,dx\,.$$

In this reformulation, we may pass to the limit $r \to 0$ also in the first term and we are left with (note that the difference between $\int_\Omega \mathbf{u}^\varepsilon \cdot \nabla(\omega_r * [\frac{|\mathbf{u}^\varepsilon|^2}{2} + p^\varepsilon]\tau)G_h\,dx$ and $+\int_\Omega \omega_r * [\mathbf{u}^\varepsilon \cdot \nabla(\frac{|\mathbf{u}^\varepsilon|^2}{2} + p^\varepsilon)]\tau G_h\,dx$ tends to zero as $r \to 0$)

$$\int_\Omega (\frac{|\mathbf{u}^\varepsilon|^2}{2} + p^\varepsilon)\delta_h\tau\,dx + \varepsilon\int_\Omega |\mathbf{u}^\varepsilon|^4\tau G_h\,dx$$
$$\leq \int_\Omega |\mathbf{f}\cdot\mathbf{u}^\varepsilon + \varepsilon\,\mathrm{div}\,|\mathbf{u}^\varepsilon|^2\mathbf{u}^\varepsilon - \mathrm{div}\,\mathbf{f}|\tau G_h\,dx$$
$$+\text{ terms containing } \nabla\tau \text{ and } \nabla^2\tau\,.$$

The terms with $\nabla\tau$ and $\nabla^2\tau$ can be estimated by integrals with integrands

$$\frac{1}{K}A_\varepsilon = |\nabla\tau||\nabla(\frac{|\mathbf{u}^\varepsilon|^2}{2} + p^\varepsilon)|G_h + |\nabla^2\tau||\frac{|\mathbf{u}^\varepsilon|^2}{2} + p^\varepsilon|G_h\,,$$
$$\frac{1}{K}B_\varepsilon = |\nabla\tau|(|\mathbf{u}^\varepsilon|^3 + |p^\varepsilon||\mathbf{u}^\varepsilon|)G_h\,,$$

K being a constant not depending on ε, h. We estimate

$$\varepsilon\,\mathrm{div}(\mathbf{u}^\varepsilon|\mathbf{u}^\varepsilon|^2)G_h\tau \leq \frac{\varepsilon}{2}\tau|\mathbf{u}^\varepsilon|^4 G_h + \varepsilon|\nabla\mathbf{u}^\varepsilon|^2 G_h\tau\,.$$

We now pass to the limit $\varepsilon \to 0$, drop the term $\frac{\varepsilon}{2}|\mathbf{u}^\varepsilon|^4 G_h\tau$ and obtain

Proposition 3.5. *Every weak cluster-point* $\mathbf{u}, p$ *of the sequence* $(\mathbf{u}^\varepsilon, p^\varepsilon)$ *satisfies*

$$\int_\Omega (\frac{\mathbf{u}^2}{2} + p)\delta_h\tau\,dx \leq \int_\Omega |\mathbf{f}\cdot\mathbf{u} - \mathrm{div}\,\mathbf{f}|\,\tau\,G_h\,dx + \limsup A_\varepsilon + B_\varepsilon\,. \tag{3.6}$$

In the next section, we shall derive uniform properties of the functions G_h. For any dimension we have uniformly in h

$$\|\nabla G_h\|_{L^{n/(n-1)-\delta}(\Omega)} + \|G_h\|_{L^{n/(n-2)-\delta}(\Omega)} \leq K \qquad (h \to 0, \|\delta_h\|_{L^1(\Omega)} \leq 1) \tag{3.7}$$

and, if V is a neighbourhood not intersecting $\operatorname{supp} \delta_h$,

$$\|G_h\|_{L^\infty(V)} \leq K \qquad (h \to 0) \tag{3.8}$$

provided that the dimension is 5. With a so called dimension-reduction technique the authors succeeded in [13] to prove (3.8) also in the case $n = 6$. There is still some hope to arrive up to $n = 7$.

Once (3.8) is known, the terms $A_\varepsilon, B_\varepsilon$ are uniformly bounded if $n = 5$ or $n = 6$, the term $\int_\Omega \mathbf{f} \cdot \nabla G_h \, dx$ is fine in any dimension for $\mathbf{f} \in L^{n+\delta}(\Omega)$, the term $\int_\Omega \mathbf{f} \cdot \mathbf{u} \, G_h \, dx$ is fine for $n = 5$, $\mathbf{f} \in L^{\frac{10}{7}+\delta}(\Omega)$. If $n = 6$, it may be treated by first proving $L^{4-\delta}$-regularity using $\delta_h = \frac{1}{(|x|^2+h^2)^2}$.

We collect our results to obtain

Proposition 3.9. *Let $n = 5$ and $\mathbf{u}, p$ the weak H^1-limit resp. $H^{n/(n-2)}$-limit of the ε-solutions $\mathbf{u}^\varepsilon, p^\varepsilon$. Let $\mathbf{f} \in L^{\frac{10}{3}}(\Omega)$. Then $\mathbf{u}, p$ solves the Navier-Stokes equations (1.1) and we have*

$$\frac{\mathbf{u}^2}{2} + p \leq K_{\Omega_0}$$

on interior domains $\Omega_0 \subset\subset \Omega$.

Remark 3.10. As mentioned, the authors were able to extend this up to six dimensions.

PROOF of Proposition 3.9: The proof follows from (3.6) and the above remarks estimating the right-hand side of (3.6). One chooses $\delta_h = \frac{1}{|B_h|}$ on $B_h(x_0)$, $\delta_h = 0$ elsewhere, and passes to the limit $h \to 0$. ■

4. The dual equation

The dual equation is of great importance for our approach.
It reads: Find $G_h \in H_0^1(\Omega) \cap L^q(\Omega)$ such that

$$\int_\Omega \nabla G_h \nabla \varphi \, dx + \int_\Omega u_i G_h D_i \varphi \, dx = \int_\Omega \delta_h \varphi \, dx \qquad \forall \varphi \in C_0^\infty(\Omega), \tag{4.1}$$

where, mostly, δ_h is a discrete Dirac function, or $\delta_h = (|x - x_0|^2 + h^2)^{-s}$.

Proposition 4.2. *For every $\delta_h \in L^\infty(\Omega)$ and $\mathbf{u} \in H_0^1(\Omega)$, $\operatorname{div} \mathbf{u} = 0$, equation (4.1) has a solution $G_h \in L^\infty(\Omega) \cap H_0^1(\Omega)$. If $\delta_h \geq 0$, then $G_h \geq 0$.*

PROOF : Approximate the term G_h by $\frac{G_{h,\delta}}{1+\delta|G_{h,\delta}|}$. The corresponding elliptic differential equation can be shown to be solvable in $H^1(\Omega)$ via a Ritz-Galerkin method. The term $G_{h,\delta}/(1 + \delta|G_{h,\delta}|)$ is a weak, completely continuous nonlinearity, furthermore $\int_\Omega u_i z/(1 + \delta|z|) D_i z \, dx = 0$ for smooth z, since $\operatorname{div} \mathbf{u} = 0$. This is applied for z being the Galerkin-approximations. Hence the corresponding nonlinear form is coercive in $H^1(\Omega)$ and the Ritz-Galerkin approximation is shown to converge. Hence we have a solution of

$$\int_\Omega \nabla G_{h,\delta} \nabla \varphi \, dx + \int_\Omega u_i G_{h,\delta}(1+\delta|G_{h,\delta}|)^{-1} D_i \varphi \, dx = \int_\Omega \delta_h \varphi \, dx \tag{4.3}$$

in $H_0^1(\Omega)$.

Proposition 4.4. *The solutions $G_{h,\delta}$ of (4.3) are uniformly bounded in $L^\infty(\Omega)$ as $\delta \to 0$.*

The proof uses the Moser technique. For this one uses the function

$$[G_{h,\delta}]_L^{s-1}, \qquad s \geq 2$$

as test functions, where $[\xi]_L = \xi$ for $|\xi| \leq L$ and $[\xi]_L = \operatorname{sign} \xi \, L$ for $|\xi| \geq L$. This truncation procedure is necessary to keep the test function in $H^1(\Omega)$. For an increasing sequence of exponents s one obtains a bound for $[G_{h,\delta}]_L^{s/2}$ in $H^1(\Omega)$ uniformly in L (s fixed) and passes to the limit $L \to \infty$. This implies that $G_{h,\delta} \in L^q(\Omega)$ $\forall q < \infty$. Note that we used

$$\int_\Omega u_i G_{h,\delta} D_i [G_{h,\delta}]_L^{s-1} \, dx = 0$$

since $\operatorname{div} \mathbf{u} = 0$. This leads to an inequality

$$\int_\Omega |\nabla (G_{h,\delta}^{s/2})|^2 \, dx \leq K \, s \int_\Omega \delta_h G_{h,\delta}^{s-1} \, dx \,. \tag{4.5}$$

Inequality (4.5) is the typical inequality where Moser's technique is applied. It leads via Sobolev's inequality to a recursion relation of the $L^{s\frac{n}{n-2}}$-norm of $G_{h,\delta}$ by the L^{s-1}-norm. By an iteration argument - in spite the factor s at the right-hand side of (4.5) - it is well known that (cf. [18]) one obtains a bound for $\|G_{h,\delta}\|_s$ which is uniformly in s and δ. Passing to the limit $s \to \infty$ we obtain

$$\|G_{h,\delta}\|_\infty \leq K_h \qquad \text{uniformly in } \delta \,,$$

and, selecting subsequences as $\delta \to 0$, we obtain a solution $G_h \in L^\infty(\Omega) \cap H_0^1(\Omega)$ of (4.1). Clearly its $L^\infty \cap H^1$-norm depends on h. Since $\mathbf{u} \in L^{\frac{2n}{n-2}}(\Omega)$ and $D_i G_h \in L^2(\Omega)$ we have $\Delta G_h \in L^{1+\sigma}(\Omega)$, $\sigma = \sigma(n)$. Thus one may work with some $L^{1+\varepsilon}$-norm of the second derivatives. The non-negativity of G_h follows by testing with $(G_h)_-$. The Propositions 4.2 and 4.4 are proved. ■

We finally need uniform estimates for G_h as $h \to 0$.

Proposition 4.6. *For all $q < \frac{n}{n-2}$ and $\tilde{q} < \frac{n}{n-1}$ we have uniformly with respect to $h \to 0$*

$$\|\nabla G_h\|_q + \|G_h\|_{\tilde{q}} \leq K_{q,\tilde{q}}\,. \tag{4.7}$$

PROOF : This follows by testing the equation with $\frac{G_h}{\sqrt[s]{1+G_h^s}}$ with $s > 0$ small ! One concludes an L^1-estimate for

$$\frac{|\nabla G_h|^2}{(1+G_h^s)^{1+1/s}}\,,$$

which implies (4.7). ∎

It remains to show the uniform L^∞-estimate for G_h outside a neighbourhood of the singularity x_0.

Proposition 4.8. *Let G_h solve (4.1), $n = 5$. Then, for every set $V \subset\subset \Omega \setminus \{x_0\}$, we have*

$$\|G_h\|_{L^\infty(V)} \leq K = K(V)$$

uniformly as $h \to 0$ and $\varepsilon \to 0$.

PROOF : It suffices to consider $V = B_R$, $B_R \subset\subset \Omega \setminus \{x_0\}$. We use the local version of Moser's technique and choose the functions

$$G_h^{s-1}\tau_s^2$$

as test functions. Here τ_s is a nonnegative Lipschitz function with support in B_{2R}, and $\tau_s \to \chi(B_R) =$ characteristic function of B_R as $s \to \infty$. The precise definition will be done later. τ_s will have the property $|\nabla\tau_s| \leq K\,s$. We obtain

$$\frac{1}{2}(s-1)\int_\Omega |\nabla G_h|^2 G_h^{s-1}\tau_s^2\,dx \leq \frac{K}{s-1}\int_\Omega |\nabla\tau_s|^2 G_h^s\,dx + \frac{2}{s}\int_\Omega u_i G_h^s D_i\tau_s\tau_s\,dx$$

and, via Sobolev's and Hölder's inequality

$$\begin{aligned}\left(\int_\Omega G_h^{3s}\tau_s^6\,dx\right)^{1/3} &\leq K\,s^2\int_{\text{supp}\,\tau_s} G_h^s\,dx + K\|\mathbf{u}\|_{\frac{10}{3}}\left(\int_{\text{supp}\,\tau_s} G^{\frac{10}{7}s}\,dx\right)^{7/10} \\ &\leq K s^2\left(\int_{\text{supp}\,\tau_s} G_h^{\frac{10}{7}s}\,dx\right)^{7/10},\end{aligned} \tag{4.9}$$

where $K = K_R$. We choose a sequence of numbers $s = s_i$ such that

$$s_0 = \frac{7}{10}\left(\frac{5}{4} - \delta\right), \qquad s_{i+1} = \frac{37}{10}\,s_i$$

and arrange that $\tau_{s_i} = 1$ on concentric balls $B^{(i)}$, $\operatorname{supp}\tau_{s_{i+1}} \subset B^{(i)}$ since $s_i = \left(\frac{21}{10}\right)^i s_0$, we can arrange that $B^{(i)} \subset B_{2R}$, $B^{(i)} \supset B_R$, $|\nabla\tau_{s_i}| \le K\,R^{-1}s_i^{-1}$. Hence (4.9) implies the recursion relation (set $l_i = \frac{7}{10}s_i$)

$$\left(\int_{B^{(i+1)}} G_h^{l_{i+1}}\,dx\right)^{1/l_{i+1}} \le (K\,l_i^\sigma)^{1/l_i}\left(\int_{B^{(i)}} G_h^{l_i}\,dx\right)^{1/l_i}$$

with some fixed $\sigma > 1$. As before, when we applied the global Moser technique, one derives a bound for $\|G_h\|_{l_{i+1}}$ on B_R as $i \to \infty$, and hence, we obtain a bound for $\|G_h\|_{L^\infty(B_R)}$. ■

5. An $L^{4-\delta}$-estimate for u

The methods of Sections 2 – 4 imply that those solutions, which are limits of ε-solutions, satisfy the inequality

$$\frac{\mathbf{u}^2}{2} + p \le K_{\Omega_0} \tag{5.1}$$

and

$$\begin{aligned} \int_{\Omega_0} \Big|\frac{\mathbf{u}^2}{2} + p\Big| \frac{1}{|x-x_0|^{n-2}}\,dx \le K\,, \\ \int_{\Omega_0} (|p| + |\mathbf{u}|^2)\frac{1}{|x-x_0|^{n-2-\delta}}\,dx \le K\,. \end{aligned} \tag{5.2}$$

This was proved here for $n = 5$; the original reference is [8] furthermore, in a new paper [13] this is extended to $n = 6$.

Proposition 5.3. *Inequality (5.2) implies $p \in L^{2-\delta}_{loc}(\Omega)$ for all $\delta > 0$ small.*

PROOF : One introduces the auxiliary equation

$$-\Delta h_L = [p]_L^{1-\delta}\,,$$

where $[p]_L = p$ if $|p| \le L$, $[p]_L = L\operatorname{sign} p$ for $|p| \ge L$, and observes that

$$\int_\Omega [p]_L^{1-\delta}\frac{1}{|x-x_0|^{n-2}}\,dx \le K$$

uniformly as $L \to \infty$, if x_0 varies on interior domains. Hence $h_L \in L^\infty_{loc}(\Omega) \cap H^1_0(\Omega)$ uniformly as $L \to \infty$ and $h_L \in H^{2,q}_{loc}(\Omega)$, $q < \infty$ for fixed L. We test the pressure equation with h_L. Since the pressure is obtained as a weak limit of the p^ε, we have

$$-\Delta p = D_i(u_k D_k u_i)$$

in a weak sense. We test this equation with τh_L, where τ is a localization function. This yields

$$\big(\nabla p, \nabla(\tau h_L)\big) = -\big(u_k D_k u_i, D_i(\tau h_L)\big)\,. \tag{5.4}$$

Since $\mathbf{u}$ is a weak H^1-limit of smooth divergence free functions $\mathbf{u}_k^m$ we have

$$\begin{aligned} -\big(u_k D_k u_i, D_i(\tau h_L)\big) &= o(1) + \big(D_i u_k^m D_k u_i^m, \tau h_L\big) \\ &\leq o(1) + K\|\tau h_L\|_{L^\infty} \qquad \text{as } m \longrightarrow \infty\,, \end{aligned}$$

where K does not depend on m since $\mathbf{u}_k^m \rightharpoonup \mathbf{u}$ weakly in $H^1(\Omega)$. Hence the right-hand side of (5.4) is estimated by

$$K\|h_L\|_{L^\infty(\Omega_0)} < \infty \qquad (L \to \infty)\,.$$

At the left-hand side of (5.4) arises the term $\int_\Omega p\tau\Delta h_L\,dx$ and terms containing $\nabla\tau$ and $\nabla^2\tau$ which are of a better order. The term $p\nabla\tau\nabla h_L$ must be rewritten as $-\nabla p\nabla\tau h_L$ plus lower order terms, in order to use the uniform L^∞-bound for h_L. All this finally yields a bound

$$\int_\Omega p\,[p]_L^{1-\delta}\tau\,dx \leq K \qquad (L \to \infty)$$

and passing to the limit $L \to \infty$ we obtain Proposition 5.2. ■

Concerning the unknown function $\mathbf{u}$ we obviously may conclude:

Proposition 5.5. *Under the assumptions of Proposition 5.3 and the one sided condition (5.1) we conclude* $\mathbf{u} \in L_{loc}^{4-4\delta}(\Omega)$.

6. The hole-filling method

The hole-filling method is a powerful tool (first used in [37]) in the theory of elliptic (and parabolic (cf. [33])) equations. It provides a simple method to gain a Morrey condition for a function. The idea is presented in the following general setting:

Proposition 6.1. *Let $z \in L^1(\Omega)$ be a function such that the <u>hole-filling inequality</u> holds*

$$\int_{B_R} |z|\,dx \leq K \int_{T_R} |z|\,dx + K_0 R^\gamma\,, \qquad T_R = B_{2R} \setminus B_R \tag{6.2}$$

for all concentric balls with radius R such that $B_{2R} \subset \Omega$, with constants $K, K_0, \gamma > 0$. Then

$$\int_{B_r} |z|\,dx \leq 2^\theta \left(\frac{r}{R_0}\right)^\theta \int_{B_{R_0}} |z|\,dx + \tilde{K}R^\gamma\,, \tag{6.3}$$

with $\theta = \log_2(1 + \frac{1}{K})$ *and a constant* $\tilde{K}$. ($\log_2$= *logarithm with basis 2).*

PROOF : Adding the term $K \int_{B_R} |z|\,dx$ to both sides of (6.2) we fill the "hole" in the torus T_R and obtain

$$(1 + K) \int_{B_R} |z|\,dx \le K \int_{B_{2R}} |z|\,dx + KR^\gamma .$$

Dividing by $1 + K$ we obtain

$$\int_{B_R} |z|\,dx \le 2^{-\theta} \int_{B_{2R}} |z|\,dx + KR^\gamma$$

and, via an iteration argument, we get (6.3) for $r = 2^{-N}R_0$ without the factor 2^θ. This factor appears to cover the cases $2^{-N+1}R_0 < r < 2^{-N}R_0$. ■

Let us explain how this method is applied in the case of the equation

$$-\Delta u = 0 .$$

(Clearly the hole filling method is not important here since there are many other ways to prove regularity.)

One uses the function

$$\zeta^2(u - \bar{u})|x - x_0|^{2-n}$$

as a test function (anxious mathematicians should use a regularization procedure replacing $|x - x_0|$ by $((x - x_0)^2 + h^2)^{\frac{1}{2}}$). Here $\bar{u}$ is the mean value of u taken over the torus $T_R = B_{2R} \setminus B_R$. The function ζ is a localization function such that $\operatorname{supp}\zeta \subset B_{2R}$, $\zeta = 1$ on B_R, $|\nabla\zeta| \le R^{-1}$, $B_R = B_R(x_0)$ etc.. We then obtain

$$\int_\Omega |\nabla u|^2 \zeta^2 |x - x_0|^{2-n}\,dx + \frac{1}{2} \int_\Omega \nabla(u - \bar{u})^2 \nabla(\zeta^2|x - x_0|^{2-n})\,dx = 0 .$$

The term

$$A_0 = \int_\Omega \nabla(u - \bar{u})^2 \nabla(\zeta^2 |x - x_0|^{2-n})\,dx$$

is rewritten as

$$A_0 = \int_\Omega \nabla((u - \bar{u})^2\zeta^2) \nabla |x - x_0|^{2-n}\,dx$$
$$+ \text{ terms containing } \nabla\zeta .$$

We use that $\nabla\zeta = 0$ on B_R and estimate the pollution terms containing $\nabla\zeta$ by

$$K \int_{T_R} |\nabla u|^2 R^{2-n}\,dx + K \int_{T_R} |u - \bar{u}|^2 R^{-n}\,dx .$$

Using Poincare's inequality we estimate $R^{-2}\int_{T_R}|u-\bar{u}|^2\,dx$ by $K\int_{T_R}|\nabla u|^2\,dx$. This yields

$$A_0=(u-\bar{u})^2(x_0)+B_0$$

$$|B_0|\le K\int_{T_R}|\nabla u|^2R^{2-n}\,dx\le K\int_{T_R}|\nabla u|^2|x-x_0|^{2-n}\,dx\,.$$

Note that we used

$$\Delta|x-x_0|^{2-n}=\delta_{x_0}\,,$$

where δ_{x_0} is Dirac's functional. Thus we obtain the hole filling inequality for $\int_\Omega|\nabla\mathbf{u}|^2|x-x_0|^{2-n}\,dx$.

In our case, we would like to arrange the hole filling inequality for the quantity $\int_{B_R}\mathbf{u}^2|x-x_0|^{2-n}\,dx$ or, say,

$$\int_{B_R}\mathbf{u}^2|x-x_0|^{2-n}+|\nabla\mathbf{u}|^2|x-x_0|^{4-n}\,dx\,.$$

To achieve this one could try to test the Navier-Stokes equations by $\mathbf{u}\,\zeta^2|x-x_0|^{4-n}$, however this is not so simple since there is the <u>trilinear</u> (in $\mathbf{u}$) term

$$u_iD_i(\frac{\mathbf{u}^2}{2}+p)\zeta^2|x-x_0|^{4-n}\,.$$

Furthermore, there is the problem that testing by $\mathbf{u}\,\zeta^2|x-x_0|^{4-n}$ or better by $\mathbf{u}\,\zeta^2\big(|x-x_0|^2+h^2\big)^{(4-n)/2}$ is not allowed since $\mathbf{u}$ is not known a priori in $L^4(\Omega)$.

The fact that it is not allowed to use $\mathbf{u}$ as a test function, a priori, has been overcome in [11] by an involved argument provided that $\frac{\mathbf{u}^2}{2}+p\le K$. One of the ideas is that $\frac{\mathbf{u}^2}{2}+p-2K$ has a sign and that it is possible to "normalize" with the factor $[1+\varepsilon(2K-\frac{\mathbf{u}^2}{2}-p)]^{-1}$. This works for any dimension. In the present course we confine ourselves to the case that $\mathbf{u}$ is the limit of ε-solutions (see Section 3) and that $n=5$. On the ε level, we may test by $\mathbf{u}^\varepsilon\zeta^2(|x-x_0|^2+h^2)^{(4-n)/2}$ since $\mathbf{u}^\varepsilon\in L^4(\Omega)$, $\nabla p^\varepsilon\in L^{4/3}(\Omega)$. The term

$$\int_\Omega u_i^\varepsilon D_i(\frac{|\mathbf{u}^\varepsilon|^2}{2}+p^\varepsilon)\zeta^2(|x-x_0|^2+h^2)^{(4-n)/2}\,dx$$

is rewritten via integration by parts:

$$\int_\Omega u_i^\varepsilon(\frac{|\mathbf{u}^\varepsilon|^2}{2}+p^\varepsilon)D_i(\zeta^2(|x-x_0|^2+h^2)^{(4-n)/2})\,dx$$

and we may pass to the limit $\varepsilon\to0$ for a subsequence since $\mathbf{u}^\varepsilon$ is uniformly bounded in $L^{\frac{10}{3}}(\Omega)$. The other terms, like $\int_\Omega|\nabla\mathbf{u}^\varepsilon|^2\zeta^2(|x-x_0|^2+h^2)^{(4-n)/2}\,dx$ are treated by a lower semicontinuity argument. Thereafter, the passage $h \to 0$ is performed and we arrive at

$$\begin{aligned}
&\int_\Omega |\nabla \mathbf{u}|^2 \zeta^2 |x-x_0|^{4-n}\,dx + \int_\Omega \mathbf{u}^2\zeta^2 \Delta |x-x_0|^{4-n}\,dx \\
&\le \int_\Omega u_i\Big(\frac{\mathbf{u}^2}{2}+p\Big) D_i(\zeta |x-x_0|^{2-n})\,dx + \int_\Omega \mathbf{f}\cdot \mathbf{u}\zeta^2 |x-x_0|^{4-n}\,dx \\
&\quad + \text{ terms with } \nabla\zeta\,.
\end{aligned}$$

The nonquadratic term is treated in the following way:
Introduce the auxiliary equation

$$-\Delta g = \frac{\mathbf{u}^2}{2} + p\,, \qquad g \in H_0^1(\Omega)\,. \tag{6.4}$$

Since $(\frac{\mathbf{u}^2}{2}+p)|x-x_0|^{2-n}$ is uniformly bounded in L^1 on compactly contained subdomains we conclude $g \in L^\infty_{loc}(\Omega)$ (indeed, test by the fundamental solution of $-\Delta$). Testing equation (6.4) by $\zeta^2|x-x_0|^{2-n}(g-\bar g)$, $\bar g$ being the mean value of g over T_R we obtain (see our introduction concerning hole filling for $\Delta u = 0$) (see also [11], [28])

$$\begin{aligned}
&\int_\Omega |\nabla g|^2 \zeta^2 |x-x_0|^{2-n}\,dx \\
&\le K\int_\Omega \Big|\frac{\mathbf{u}^2}{2}+p\Big| |x-x_0|^{2-n}\zeta^2\,dx + \int_{T_R} |\nabla g|^2 |x-x_0|^{2-n}\,dx\,.
\end{aligned}$$

We apply the last inequality for estimating

$$\begin{aligned}
&\Big|\int_\Omega u_i\Big(\frac{\mathbf{u}^2}{2}+p\Big) D_i\big(\zeta^2|x-x_0|^{4-n}\big)\,dx\Big| = \Big|\int_\Omega u_i \Delta g D_i\big(\zeta^2|x-x_0|^{4-n}\big)\,dx\Big| \\
&\le \Big|\int_\Omega \nabla u_i \nabla g D_i\big(\zeta^2 |x-x_0|^{4-n}\big)\,dx\Big| + \Big|\int_\Omega u_i \nabla g \nabla D_i\big(\zeta^2|x-x_0|^{4-n}\big)\,dx\Big| \\
&\le \varepsilon_0 \int_{B_{2R}} |\nabla \mathbf{u}|^2 |x-x_0|^{4-n}\,dx + \frac{K}{\varepsilon}\int_{B_{2R}} |\nabla g|^2 |x-x_0|^{2-n}\,dx \\
&\quad + \varepsilon_0 \int_{B_{2R}} |\mathbf{u}|^2 |x-x_0|^{2-n}\,dx \\
&\le \varepsilon_0 \int_{B_{2R}} \big\{|\nabla \mathbf{u}|^2|x-x_0|^{4-n} + |\mathbf{u}|^2|x-x_0|^{2-n}\big\}\,dx \\
&\quad + \frac{K}{\varepsilon}\int_{B_{2R}} \Big|\frac{\mathbf{u}^2}{2}+p\Big| |x-x_0|^{2-n}\,dx + \frac{K}{\varepsilon}\int_{T_R} |\nabla g|^2 |x-x_0|^{2-n}\,dx\,.
\end{aligned}$$

Hence we arrive at

$$\begin{aligned}
&\int_{B_R} |\nabla \mathbf{u}|^2 |x-x_0|^{4-n}\,dx + \int_{B_R} |\mathbf{u}|^2 |x-x_0|^{2-n}\,dx \\
&\quad + \int_{B_R} |\nabla g|^2 |x-x_0|^{2-n}\,dx \\
&\le K \int_{T_R} |\nabla \mathbf{u}|^2 |x-x_0|^{4-n}\,dx + K \int_{T_R} |\mathbf{u}|^2 |x-x_0|^{2-n}\,dx \\
&\quad + K \int_{T_R} |\nabla g|^2 |x-x_0|^{2-n}\,dx + K \int_{B_{2R}} \Big|\frac{\mathbf{u}^2}{2} + p\Big| |x-x_0|^{2-n}\,dx + KR^\alpha ,
\end{aligned} \tag{6.5}$$

where the term KR^α comes from the exterior force $\mathbf{f}$. We see that the hole-filling paradise is not yet reached since there is the term

$$K \int_{B_{2R}} \Big|\frac{\mathbf{u}^2}{2} + p\Big| |x-x_0|^{2-n}\,dx .$$

This term is estimated by using the pressure equation similar as in Section 2, however in a localized version (test the pressure equation by $\zeta^2(|x-x_0|^2+h^2)^{\frac{4-n}{2}}$, pass to the limit $h \to 0$, use that $\int_\Omega \left(\frac{\mathbf{u}^2}{2}+p\right)_+ |x-x_0|^{2-n}\,dx \le K$). This leads to an inequality

$$\begin{aligned}
\int_{B_R} \Big|\frac{\mathbf{u}^2}{2} + p\Big| |x-x_0|^{2-n}\,dx &\le K \int_{T_R} \Big|\frac{\mathbf{u}^2}{2} + p\Big| |x-x_0|^{2-n}\,dx \\
&\quad + \int_{T_R} \mathbf{u}^2 |x-x_0|^{2-n}\,dx + KR^\alpha .
\end{aligned} \tag{6.6}$$

Adding (6.6), multiplied by a large factor Λ, to (6.5) we obtain the hole-filling inequality for the function

$$\mathbf{u}^2|x-x_0|^{2-n} + |\nabla \mathbf{u}|^2|x-x_0|^{4-n} + |\nabla g|^2|x-x_0|^{2-n} + \Lambda\Big|\frac{\mathbf{u}^2}{2} + p\Big||x-x_0|^{2-n} .$$

Thus we obtain

Theorem 6.7. *Let $n=5$, $\mathbf{f} \in L^{\frac{n}{2}+\delta}, \delta > 0$, and let $\mathbf{u}$ be a solution of the stationary Navier-Stokes equations, which is the limit of ε-solutions. Then*

$$\int_{B_R} \mathbf{u}^2|x-x_0|^{2-n} + |\nabla \mathbf{u}|^2|x-x_0|^{4-n}\,dx \le KR^\alpha$$

for some $\alpha > 0$, uniformly with respect to $x_0 \in \Omega_0 \subset\subset \Omega$, $0 < R \le R_0$.

With refined techniques, the authors were also able to cover the case $n = 6$, and in the periodic case up to $n = 15$. No restriction of the dimension is necessary once $\frac{\mathbf{u}^2}{2} + p \le K$ is known.

7. The bootstrap argument in Morrey spaces

This section contains the final step towards regularity of solutions of the stationary Navier-Stokes equations. We use a new precise version of the imbedding between Morrey spaces and recent results in the linear theory of the Stokes and Laplace equation (cf. [31], [4]). We first recall the definition of the Morrey space $L^{q,\lambda}(\Omega)$ for $1 \le q < \infty$ and $0 \le \lambda \le n$.

Definition 7.1. *The space $L^{q,\lambda}(\Omega)$ consists of all functions z such that*

$$\sup_{B_R} R^{\lambda-n} \int_{B_R} |z|^q \, dx < \infty ,$$

where the supremum is taken over all balls $B_R \subset \mathbb{R}^n$ (The function z is assumed to be zero on $\mathbb{R}^n \setminus \Omega$).

The local analog $L^{q,\lambda}_{loc}(\Omega)$ is defined by requiring that the corresponding supremum is finite under the restriction $B_R \subset \Omega_0$, for all $\Omega_0 \subset\subset \Omega$. One has the following imbedding theorem

Theorem 7.2. *Let $1 < q < n$, $0 \le \lambda \le n$ and $\frac{1}{q} - \frac{1}{\lambda} > 0$. Let $\nabla z \in L^{q,\lambda}_{loc}(\Omega)$. Then*

$$z \in L^{q^*,\lambda}_{loc}(\Omega) , \qquad \frac{1}{q^*} = \frac{1}{q} - \frac{1}{\lambda} .$$

PROOF : [31], [4] ■

Note that the Morrey exponent λ is preserved (it is quite convenient not to "loose a δ") and that, for $\lambda = n$, one has the usual Sobolev imbedding theorem.

Furthermore we shall use the following

Theorem 7.3. *Let $\mathbf{u} \in H^1(\Omega)$ satisfy*

$$\begin{aligned} -\Delta \mathbf{u} + \nabla p &= \mathbf{f} \\ \operatorname{div} \mathbf{u} &= 0 \end{aligned} \qquad \text{in } \Omega ,$$

with $\mathbf{f} \in L^{q,\lambda}_{loc}(\Omega)$ for some $q \in (1,n)$, $\lambda \in [0,n]$. Then

$$\nabla^2 \mathbf{u} \in L^{q,\lambda}_{loc}(\Omega) .$$

PROOF : [4] ■

A similar theorem holds just for $-\Delta \mathbf{u} = \mathbf{f}$.

From Section 6 we know that $\nabla \mathbf{u} \in L^{2,4-\delta}_{\mathrm{loc}}(\Omega)$, hence, by Theorem 7.2

$$\mathbf{u} \in L^{q_0,4-\delta}_{\mathrm{loc}}(\Omega)$$

where $\frac{1}{q_0} = \frac{1}{2} - \frac{1}{4-\delta}$. By Hölder's inequality

$$\mathbf{u} \cdot \nabla \mathbf{u} \in L_{\mathrm{loc}}^{s,4-\delta}(\Omega)$$

with $\frac{1}{s} = \frac{1}{q_0} + \frac{1}{2}$. For simplicity we assume $\mathbf{f} \in L^{\infty}(\Omega)$ and do not work out the case $\mathbf{f} \in L^q(\Omega)$. Using Theorem 7.3 we conclude

$$\nabla^2 \mathbf{u} \in L_{\mathrm{loc}}^{s,4-\delta}(\Omega)$$

and using Theorem 7.2 twice (conclude first that $\nabla \mathbf{u}$ is in some Morrey space and then $\mathbf{u}$) we obtain

$$\mathbf{u} \in L_{\mathrm{loc}}^{q_1,4-\delta}(\Omega)$$

with

$$\frac{1}{q_1} = \frac{1}{s} - \frac{2}{4-\delta} = \frac{1}{q_0} + \frac{1}{2} - \frac{2}{4-\delta}$$

i. e.

$$\frac{1}{q_1} = \frac{1}{q_0} - \sigma \qquad \text{where } \sigma = \frac{2}{4-\delta} - \frac{1}{2} > 0\,.$$

Iterating this argument, we obtain

$$\mathbf{u} \in L_{\mathrm{loc}}^{q_i,4-\delta}(\Omega) \qquad \text{with } \frac{1}{q_i} = \frac{1}{q_0} - \sigma\,.$$

This works for all i such that $\frac{1}{q_0} - i\sigma > 0$; the maximal index i satisfies $\frac{1}{q_0} - (i+1)\sigma < 0$ which means

$$\frac{1}{q_i} - \sigma < 0 \qquad \text{or} \qquad q_i > \frac{1}{\sigma}$$

since $\sigma = \frac{2}{4-\delta} - \frac{1}{2}$ is small for small δ, q_i will be large, say larger then the dimension n, and hence the classical regularity theory can be applied. Of course, one does not need the classical theory; one can continue to work out the "Morrey-space-bootstrap" for $\nabla \mathbf{u}$ and $\nabla^2 \mathbf{u}$ similar as it has been done for $\mathbf{u}$ above.

If one wants to avoid Theorem 7.2 one has the possibility just to use a similar one only for Laplace's equation. For this one has to work out the bootstrap argument also for the pressure p via the pressure equation. The $L^{q,\lambda}$ inclusion for p with an increasing sequence of exponents q is proved by using the fact, that $\nabla\ \mathbf{u} \in L_{\mathrm{loc}}^{t,4-\lambda}(\Omega)$ for an increasing sequence of numbers t. (In the above proof concerning the function $\mathbf{u}$ itself we worked with $t = 2$). Thus we have obtained:

Theorem 7.4. *Let* $\mathbf{u}$ *be a solution to the stationary Navier-Stokes equations in* n*-dimensions, let* $\mathbf{f} \in L_{\mathrm{loc}}^q(\Omega)$, $1 \leq q < \infty$ *and let* $\mathbf{u} \in L_{\mathrm{loc}}^{2,2-\delta}(\Omega)$ *and* $\nabla \mathbf{u} \in L_{\mathrm{loc}}^{2,4-\delta}(\Omega)$ *for some* $\delta > 0$. *Then* $\mathbf{u} \in H_{\mathrm{loc}}^{2,q}(\Omega)$.

Recall that, for $n = 5$ (and $n = 6$ cf. [13] and $n \leq 15$ cf. [12]) we indeed have constructed solutions satisfying these assumptions.

8. Solutions with fast decay at infinity

G. P. Galdi suggested to try to apply the preceding methods (we would like to call them "head pressure techniques") to the stationary Navier-Stokes equations in $\mathbb{R}^3$, in order to obtain a solution $\mathbf{u}$ in $\mathbb{R}^3$, such that

$$\int_{\mathbb{R}^3} |\mathbf{u}|^2 |x|^{-1-\delta}\, dx \le K\,. \tag{8.1}$$

Note that

$$\int_{\mathbb{R}^3} |\mathbf{u}|^2 |x|^{-2}\, dx \le K \int_{\mathbb{R}^3} |\nabla \mathbf{u}|^2\, dx \tag{8.2}$$

holds anyhow for all functions in $\tilde{H}^1(\mathbb{R}^3)$ = closure of C_0^∞-functions with respect to $(\int_{\mathbb{R}^3} |\nabla \cdot|^2\, dx)^{\frac{1}{2}}$. Thus property (8.1) is more than the mere property to be contained in $\tilde{H}^1(\mathbb{R}^3)$. We call it "fast decay at infinity". We refer the reader to [25], [24], [6], [7], [1], [15] and [16] for a detailed discussion of the considered problem in the case of an exterior domain and to [25], [24], [22] and [26] for the case of the whole space $\mathbb{R}^3$.

In fact we indicate in the following that Galdi's conjecture is true and that one can construct a solution in $\mathbb{R}^3$ without smallness condition on the data such that (8.1) holds for all $\delta > 0$. We consider the usual system

$$\left.\begin{aligned} -\Delta \mathbf{u} + \mathbf{u}\cdot\nabla\mathbf{u} + \nabla p &= \mathbf{f} \\ \operatorname{div}\mathbf{u} &= 0 \end{aligned}\right\} \quad \text{in } \mathbb{R}^3 \tag{8.3}$$

and assume, for simplicity, that

$$\mathbf{f} \in L^\infty(\mathbb{R}^3) \text{ and } \mathbf{f} \text{ has compact support in } \mathbb{R}^3\,. \tag{8.4}$$

The assumptions on $\mathbf{f}$ can be relaxed considerably.

Theorem 8.5. *Let* $\mathbf{f}$ *satisfy (8.4). Then there exists a solution* $\mathbf{u} \in \tilde{H}^1(\mathbb{R}^3)$ *of (8.3) such that*

$$\begin{aligned} &\int_{\mathbb{R}^3} |\mathbf{u}|^2 |x|^{-1-\delta}\, dx < \infty \qquad \text{for all } \delta > 0 \\ &\int_{\mathbb{R}^3} \Big|\frac{\mathbf{u}^2}{2} + p\Big| |x|^{-1}\, dx < \infty \\ &\int_{\mathbb{R}^3} (\mathbf{u}\cdot x)^2 |x|^{-3}\, dx < \infty\,. \end{aligned} \tag{8.6}$$

We prove Theorem 8.5 using estimates similar to the preceding local ones, combined with additional new techniques. However, the difficulty here is the analysis for

x near infinity. We would like to mention that our approach also works in the case of an exterior domain. More details, generalizations and a more advanced aspect are planned to be included in a paper with Galdi.

The construction of a solution $\mathbf{u}$ stated in Theorem 8.5 starts with the approximate system

$$\begin{aligned} -\Delta \mathbf{u} + \mathbf{u} \cdot \nabla \mathbf{u} + \alpha \mathbf{u} + \nabla p = \mathbf{f} \\ \operatorname{div} \mathbf{u} = 0 \end{aligned} \qquad \text{in } \mathbb{R}^3 , \tag{8.7}$$

where $\mathbf{u} = \mathbf{u}_\alpha$, $p = p_\alpha$, $\alpha > 0$. For $\mathbf{f} \in L^\infty(\mathbb{R}^3)$, $\mathbf{f}$ having compact support, say $\operatorname{supp} \mathbf{f} \subset B_K$, it is a simple exercise (via a Ritz-Galerkin method) to prove that (8.7) has a weak solution, satisfying the usual weak formulation with divergence free smooth test functions, such that

$$\mathbf{u}_\alpha \in H^1(\mathbb{R}^3) , \qquad \mathbf{u}_\alpha \in W^{2,q}_{\text{loc}}(\mathbb{R}^3) \qquad \forall q < \infty ,$$

and that

$$\int_{\mathbb{R}^3} |\nabla \mathbf{u}_\alpha|^2 \, dx + \alpha \int_{\mathbb{R}^3} |\mathbf{u}_\alpha|^2 \, dx \le K$$

uniformly as $\alpha \to 0$. The approximate pressure $p = p_\alpha$ is defined as

$$p = R_i R_j (u_i u_j) - \mathbf{q} * \mathbf{f} , \tag{8.8}$$

where R_i is the Riesz potential (cf. [32]) and $\mathbf{q}$ is the volume potential for the pressure of the Stokes equation (cf. [24]), and solves

$$-\Delta p = (\nabla \mathbf{u}) \cdot (\nabla \mathbf{u})^T - \operatorname{div} \mathbf{f} . \tag{8.9}$$

We have uniformly as $\alpha \to 0$

$$p \in L^3(\mathbb{R}^3) \cap W^{1,q}_{\text{loc}}(\mathbb{R}^3) \qquad \forall q < \infty$$

and, for fixed $\alpha > 0$,

$$p \in L^{1+\delta}(\mathbb{R}^3) \qquad \forall \delta > 0 \, , \delta \text{ small.}$$

In fact, this follows from the representation (8.8) of p_α. One easily checks (using $\mathbf{u} \in L^6(\mathbb{R}^3)$ and $p \in L^3(\mathbb{R}^3)$) that (8.7) is satisfied almost everywhere (cf. [27]).

From (8.7) and (8.9) we derive the approximate <u>head pressure</u> equation

$$-\Delta\Big(\frac{\mathbf{u}^2}{2} + p\Big) + u_i D_i \Big(\frac{\mathbf{u}^2}{2} + p\Big) + \alpha \mathbf{u}^2 = (\nabla \mathbf{u}) \cdot (\nabla \mathbf{u})^T - |\nabla \mathbf{u}|^2 + \mathbf{f} \cdot \mathbf{u} - \operatorname{div} \mathbf{f} . \tag{8.10}$$

In order to obtain a one sided estimate, we solve the <u>dual</u> equation for $G = G_{\alpha,\rho} \in \tilde{H}^1(\mathbb{R}^3 \setminus B_1)$, $(B_1 = B_1(0))$,

$$\begin{aligned} -\Delta G - u_i D_i G &= \chi_\rho \chi |x|^{-1} && \text{in } \mathbb{R}^3 \setminus B_1 \\ G &= 0 && \text{on } \partial B_1 . \end{aligned} \tag{8.11}$$

Here $\chi = \operatorname{sign}\big(\frac{\mathbf{u}^2}{2} + p\big)_+$, where $\operatorname{sign} 0 = 0$, and χ_ρ is the characteristic function of the set $B_\rho(0)$.

Later, we shall pass to the limit $\rho \to \infty$. For fixed ρ, it is simple to see (via Ritz-Galerkin) that (8.11) has a solution G with $\nabla G \in L^2(\mathbb{R}^3 \setminus B_1)$, $G \in L^6(\mathbb{R}^3 \setminus B_1) \cap W^{2,q}_{\mathrm{loc}}(\mathbb{R}^3 \setminus B_1)$ for all $q < \infty$, <u>and</u> $G \geq 0$. (Note that all estimates depend on ρ and that the $W^{2,q}_{\mathrm{loc}}$-estimates hold near ∂B_1.)

The difficulty is to obtain a bound for G as $\rho \to \infty$. This is one of the reasons why we consider the dual equation on $\mathbb{R}^3 \setminus B_1$ rather than on $\mathbb{R}^3$. Let $\tau_1 = 1$ on B_1, $0 \leq \tau_1 \leq 1$, $|\nabla \tau_1| \leq 1$, τ_1 being smooth and having compact support. We test the head pressure equation (8.10) by

$$\tau_R(1-\tau_1)\, G\,,$$

where $\tau_R(x) = \tau_1(x/R)$, and perform partial integration. This yields

$$\begin{aligned} &\int_{\mathbb{R}^3} \Big(\frac{\mathbf{u}^2}{2} + p\Big)\frac{\chi_\rho \chi}{|x|} \tau_R (1-\tau_1)\, dx + \alpha \int_{\mathbb{R}^3} \mathbf{u}^2 G \tau_R (1-\tau_1)\, dx \\ &\leq A_R + K_0 \left(\int_{B_K \setminus B_1} (G^2 + |\nabla G|^2)\, dx \right)^{\frac{1}{2}} . \end{aligned} \tag{8.12}$$

Here B_K is a ball of radius K, recall $\operatorname{supp} \mathbf{f} \subset B_K$. The second term collects terms containing the factors $\nabla \tau_1$, $\Delta \tau_1$, $\mathbf{f} \cdot \mathbf{u}$, $\operatorname{div} \mathbf{f}$, note that $\mathbf{f}$ is supposed to have compact support; K_0 and K are <u>independent</u> of R, ρ, α. (The $W^{2,q}_{\mathrm{loc}}$-regularity of $\mathbf{u}$ is uniform as $\alpha \to 0$.) The terms creating A_R are

$$\Big(\frac{\mathbf{u}^2}{2} + p\Big) \operatorname{div}(\nabla \tau_R\, G)(1-\tau_1)\,, \qquad u_i\Big(\frac{\mathbf{u}^2}{2} + p\Big) D_i(\tau_R\, G)(1-\tau_1)\,. \tag{8.13}$$

Since $\mathbf{u} \in L^2(\mathbb{R}^3) \cap L^6(\mathbb{R}^3)$, $p \in L^{1+\delta}(\mathbb{R}^3) \cap L^3(\mathbb{R}^3)$, $G \in L^6(\mathbb{R}^3 \setminus B_1)$, $|\nabla \tau_R| \leq R^{-1}$ it is easy to see that the terms (8.13) tend to zero in $L^1(\mathbb{R}^3)$ as $R \to \infty$, ρ, α fixed. Thus we pass in (8.12) to the limit $R \to \infty$ and arrive at

$$\int_{\mathbb{R}^3} \Big(\frac{\mathbf{u}^2}{2} + p\Big)_+ (1-\tau_1) \frac{\chi_\rho}{|x|}\, dx \leq K_0 \left(\int_{B_K \setminus B_1} (G^2 + |\nabla G|^2)\, dx \right)^{\frac{1}{2}} . \tag{8.14}$$

We want to obtain a bound for

$$\int_{B_K \setminus B_1} (G^2 + |\nabla G|^2)\, dx\,, \qquad \text{uniformly for } \rho \to \infty, \alpha \to 0\,.$$

For this we multiply equation (8.11) by 1 and integrate over $B_r \setminus B_1$, $r > 1$. We obtain after partial integration

$$-\int_{\partial B_r} \frac{\partial G}{\partial \nu}\, do - \int_{\partial B_1} \frac{\partial G}{\partial \nu}\, do - \int_{\partial B_r} u_i \nu_i \, G \, do = \int_{B_r \setminus B_1} \frac{\chi_\rho \chi}{|x|}\, dx \tag{8.15}$$

hence

$$\Big| \int_{\partial B_r} \frac{\partial G}{\partial \nu}\, do \Big| \le k(r) \int_{\partial B_r} G \, do + K(r) + \Big| \int_{\partial B_1} \frac{\partial G}{\partial \nu}\, do \Big| \tag{8.16}$$

with smooth, say continuous functions $k(\cdot), K(\cdot) : [1, \infty) \to \mathbb{R}_+$. This follows from the fact that $\mathbf{u}$ is locally regular independent on α but dependent on r. Estimate (8.16) is uniform in ρ and α.

The term $\int_{\partial B_1} \frac{\partial G}{\partial \nu}\, do$ is estimated with a special trick, using the bidual equation

$$\begin{aligned} -\Delta z + u_i D_i z &= 0 && \text{in } B_r \setminus B_1 \,, \\ z &= 0 && \text{on } \partial B_r \,, \\ z &= 1 && \text{on } \partial B_1 \,. \end{aligned} \tag{8.17}$$

By the maximum principle, $0 \le z \le 1$. Clearly z exists, and there is the estimate $\|\nabla z\|_\infty \le k(r)$ due to regularity theory. We test the equation for G with z and obtain

$$\big|(-\Delta G, z) + (u_i D_i G, z)\big| = \Big(\frac{\chi_\rho \chi}{|x|}, z\Big) \le c\, r^2 \,. \tag{8.18}$$

Here the parentheses denote the L^2-scalar product over $B_r \setminus B_1$. By the integration by parts, (8.18) and (8.17)

$$\Big| - \int_{\partial B_1} \frac{\partial G}{\partial \nu}\, do + \int_{\partial B_r} \frac{\partial z}{\partial \nu}\, G \, do \Big| \le c\, r^2$$

due to the boundary conditions for z and G. Since $|\nabla z| \le k(r)$ on $B_r \setminus B_1$ (regularity theory) we obtain

$$\Big| \int_{\partial B_1} \frac{\partial G}{\partial \nu}\, do \Big| \le c\, r^2 + k(r) \int_{\partial B_r} G \, do$$

and finally, using also,

$$\Big| \int_{\partial B_r} \frac{\partial G}{\partial \nu}\, do \Big| \ge \frac{d}{dr} \Big(\int_{\partial B_r} G \, do \Big) - k(r) \int_{\partial B_r} G \, do$$

we arrive at

$$\frac{d}{dr} \int_{\partial B_r} G \, do \le k(r) \int_{\partial B_r} G \, do + K(r) \,. \tag{8.19}$$

From (8.19) we obtain a bound

$$\int_{\partial B_r} G\, do \le \tilde{K}(r)\,,$$

and hence

$$\int_1^r \int_{\partial B_r} G\, do\, ds \le K_1 \qquad \text{for all } r \le 4K\,.$$

Once we have this local $L^1(B_{4K} \setminus B_1)$-bound for G we get local norms of higher derivatives of G. (Start the argument by testing the equation for G with $\frac{G}{\sqrt[s]{1+G^s}}\tau_{2K}$ cf. [9, Lemma 1.45] for a similar argument.)

Hence we proved (having performed the limiting process $\rho \to \infty$ in (8.14)), still with fixed α, that

$$\int_{\mathbb{R}^3} \Big(\frac{\mathbf{u}^2}{2} + p\Big)_+ \frac{1}{|x|}\, dx \le K_0 \qquad (\text{ uniformly as } \alpha \to 0)\,. \tag{8.20}$$

(The term $\int_{\mathbb{R}^3}(\frac{\mathbf{u}^2}{2} + p)_+ \frac{\tau_1}{|x|}\, dx$ also occurring in (8.14) is bounded due to local estimates.)

We now test the pressure equation ($\alpha > 0$ fixed) by $\tau_R|x|$. There arise terms like $p\,\nabla\tau_R\,\nabla|x|$ etc., which tend to zero as $R \to \infty$ since $p \in L^{1+\delta}(\mathbb{R}^3)$. We end up (see the calculations of Section 2 or cf. [8])

$$-\int_{\mathbb{R}^3} \Big(\frac{\mathbf{u}^2}{2} + p\Big)\frac{1}{|x|}\, dx + C_0 \int_{\mathbb{R}^3} \frac{(\mathbf{u}\cdot x)^2}{|x|^3}\, dx \le K\,,$$

where $C_0 > 0$, and via (8.20)

$$\int_{\mathbb{R}^3} \Big|\frac{\mathbf{u}^2}{2} + p\Big|\frac{1}{|x|}\, dx + C_0 \int_{\mathbb{R}^3} \frac{(\mathbf{u}\cdot x)^2}{|x|^3}\, dx \le K\,. \tag{8.21}$$

This proves the first two inequalities of the theorem (passing to the limit $\alpha \to 0$ using Fatou's Lemma and $\mathbf{u}_\alpha \to \mathbf{u}, p_\alpha \to p$). For obtaining the bound

$$\int_{\mathbb{R}^3} \mathbf{u}^2 |x|^{-1-\delta}\, dx$$

a method dual to Section 2 is used: We test the pressure equation by $\tau_R|x|^{1-\delta}$. There arise terms

$$\Big(\frac{\mathbf{u}^2}{2} + p\Big)\frac{1}{|x|^{1+\delta}}\,, \qquad \frac{(\mathbf{u}\cdot x)^2}{|x|^{3+\delta}} \tag{8.22}$$

and with an <u>opposite</u> sign $\frac{u^2}{|x|^{1+\delta}}$. Because of (8.21), also the terms (8.22) are bounded in $L^1(\mathbb{R}^3)$. (Recall that the bound in a neighbourhood of 0 is no problem due to local regularity estimates.) Hence we obtain also the first inequality of (8.6) and the theorem is proved.

Theorem 8.23. *Let* $\mathbf{u}$ *be a solution of (8.3) with the property that*

$$\int_{\mathbb{R}^3} \mathbf{u}^2 |x|^{-1-\delta}\, dx < \infty\,. \tag{8.24}$$

Then this solution also satisfies

$$\int_{\mathbb{R}^3} |\nabla \mathbf{u}|^2 |x|^{1-\delta}\, dx < \infty \tag{8.25}$$

and

$$|\mathbf{u}(x)| \leq K|x|^{-1+\delta} \qquad (|x| \to \infty) \tag{8.26}$$

for all $\delta > 0$.

Estimate (8.26) follows via the representation of $\mathbf{u}$ with the fundamental solution of the linear Stokes operator and the inequalities (8.24), (8.25). Concerning (8.25) it is easy to see the weaker property

$$\int_{\mathbb{R}^3} |\nabla \mathbf{u}|^2 |x|^{1/4-\delta} dx < \infty\,. \tag{8.27}$$

This follows by testing the Navier-Stokes equations with $\mathbf{u}|x|^{1/4-\delta}\tau_R$, τ_R as before, and estimating the convective term

$$\begin{aligned} |\int_{\mathbb{R}^3} u_i D_i u_k u_k |x|^{1/4-\delta} \tau_R\, dx| &\leq K \int_{\mathbb{R}^3} |\mathbf{u}|^3 |x|^{-3/4-\delta}\, dx \\ &= K \int_{\mathbb{R}^3} |\mathbf{u}|^{3/2} |x|^{-3/4-\delta} |\mathbf{u}|^{3/2}\, dx \\ &\leq \left(\int_{\mathbb{R}^3} |\mathbf{u}|^2 |x|^{-1-\delta}\, dx \right)^{3/4} \left(\int_{\mathbb{R}^3} |\mathbf{u}|^6\, dx \right)^{1/4}. \end{aligned}$$

The term containing $\mathbf{u} \cdot \nabla p$ is estimated in a similar manner, using $p \in L^3(\mathbb{R}^3)$ and $p|x|^{-1-\delta} \in L^1(\mathbb{R}^3)$. The stronger estimate (8.25) follows by testing the Navier-Stokes equations with $\mathbf{u}|x|^{-1-\delta}\tau_R^4$, using the Sobolev inequality

$$K \int_{\mathbb{R}^3} |\nabla \left(\mathbf{u}|x|^{1/2-\delta/2} \tau_R^2 \right)|^2\, dx \geq \left(\int_{\mathbb{R}^3} \mathbf{u}^6 |x|^{3-3\delta} \tau_R^6\, dx \right)^{1/3} \tag{8.28}$$

and a related estimate for p which we are able to achieve. The right-hand side of (8.28) is used to dominate the convective and pressure term.

Bibliography

[1] K.I. Babenko, *On stationary solutions of the problem of flow past a body of a viscous incompressible fluid*, Mat. SSSR Sbornik **20** (1973), 1-25.

[2] L. Caffarelli, R. Kohn, L. Nirenberg, *Partial regularity of suitable weak solutions of the Navier-Stokes equations*, Comm. on Pure and Appl. Math. **35** (1985), 771–831.

[3] L. Cattabriga, *Su un problema al contorno relativo al sistema di equazioni di Stokes*, Rend. Sem. Mat. Padova **31** (1961), 308–340.

[4] F. Chiarenza, M. Frasca, *Morrey spaces and Hardy-Littlewood maximal function*, Rend. Mat. Appl. **7(7)** (1987), 273–279.

[5] P. Constantin, *Remarks on the Navier-Stokes equations*, New Perspectives in Turbulence, L.Sirovich (ed.), Springer, New York, 1990, pp. 229–261.

[6] R. Finn, *Estimates at infinity for stationary solutions of the Navier-Stokes equations*, Bull. Math. Soc. Sci. Math. Phys. R.P. Roumaine **3 (51)** (1959), 387–418.

[7] R. Finn, *On the exterior stationary problem for the Navier-Stokes equations, and associated problems*, Arch. Rat. Mech. Anal. **19** (1965), 363–406.

[8] J. Frehse, M. Růžička, *On the Regularity of the Stationary Navier-Stokes Equations*, Ann. Scu. Norm. Pisa **21** (1994), 63–95.

[9] J. Frehse, M. Růžička, *Weighted Estimates for the Stationary Navier-Stokes Equations*, Acta Appl. Mathematicae **37** (1994), 53–66.

[10] J. Frehse, M. Růžička, *Existence of Regular Solutions to the Stationary Navier-Stokes Equations*, Math. Ann. **302** (1995), 699–717.

[11] J. Frehse, M. Růžička, *Regularity for the Stationary Navier-Stokes Equations in Bounded Domains*, Arch. Rat. Mech. Anal. **128** (1994), 361–381.

[12] J. Frehse, M. Růžička, *Regular Solutions to the Steady Navier-Stokes Equations*, Proceedings of the 3. International Conference on Navier–Stokes Equations and Related Nonlinear Problems, Madeira 1994, A. Sequeira (ed.), Plenum Press, New York, 1995, pp. 131–139.

[13] J. Frehse, M. Růžička, *Existence of Regular Solutions to the Steady Navier-Stokes Equations in Bounded Six-Dimensional Domains*, (accepted), Ann. Scu. Norm. Pisa.

[14] J. Frehse, M. Růžička, *A New Regularity Criterion for Steady Navier-Stokes Equations*, (submitted), Diff. Int. Eq..

[15] G.P. Galdi, *On the assymptotic properties of Leray's solution to the exterior stationary three-dimensional Navier-Stokes equations with zero velocity at in-*

finity, Degenerate Diffusions, IMA Vol. in Math. and its Appl. Vol. 47, W.M. Ni, L.A. Veletier, J.L. Vasquez (eds.), Springer, 1992, pp. 95–103.

[16] G.P. Galdi, *An Introduction to the mathematical theory of the Navier-Stokes equations*, vol. II, Springer, New York, 1994.

[17] C. Gerhardt, *Stationary solutions to the Navier-Stokes equations in dimension four*, Math. Zeit. **165** (1979), 193–197.

[18] M. Giaquinta, *Introduction to regularity theory for nonlinear elliptic systems*, Birkhäuser, Basel, 1993.

[19] M. Giaquinta, G. Modica, *Nonlinear systems of the type of the stationary Navier-Stokes system*, J. Reine Angew. Math. **330** (1982), 173–214.

[20] Y. Giga, H. Sohr, *Abstract L^p Estimates for the Cauchy Problem with Applications to the Navier-Stokes Equations in Exterior Domains*, J. Funct. Anal. **102** (1991), 72–94.

[21] K.K. Golovkin, O. A. Ladyzhenskaya, *Solutions of non-stationary boundary value problems for Navier-Stokes equations*, Trudy Mat. Inst. Steklov **59** (1960), 100–114.

[22] H. Kozono, H. Sohr, *On stationary Navier-Stokes equations in unbounded domains*, Richerche di Mat. **152** (1993), 69–86.

[23] O. A. Ladyzhenskaya, *Investigation of the Navier-Stokes equation for a stationary flow of an incompressible fluid*, Uspechi Mat. Nauk **14 (3)** (1959), 75–97.

[24] O. A. Ladyzhenskaya, *The mathematical theory of viscous incompressible flow*, 2nd ed., Gordon and Breach, 1969.

[25] J. Leray, *Etudes de diverses équations intégrales non linéaires et des quelques problemes que pose l'hydrodynamique*, J. Math. Pures Appl. **12** (1933), 1–82.

[26] T. Miyakawa, *Hardy spaces of solenoidal vector fields, with applications to the Navier-Stokes equations*, preprint.

[27] J. Nečas, M. Růžička, V. Šverák, *On self-similar solutions of the Navier-Stokes equations*, (accepted), Acta Math..

[28] M. Růžička, J.Frehse, *Regularity for Steady Solutions of the Navier-Stokes Equations*, (to appear), Proceedings Oberwolfach 1994.

[29] H. Sohr, *Zur Regularitätstheorie der instationären Gleichungen von Navier-Stokes*, Math. Zeit. **184** (1983), 359–376.

[30] V.A. Solonnikov, *On estimates of the Green tensor for some boundary value problems*, (in Russian), Dokl. Akad. Nauk SSSR **130** (1960), 988–991.

[31] G.S. Stampacchia, *The spaces $L^{p,\lambda}$, $N^{p,\lambda}$ and interpolation*, Ann. Scu. Norm. Sup. Pisa **19** (1965), 443–462.

[32] E. Stein, *Singular integrals and the differentiability properties of functions*, Princeton University Press, 1970.

[33] M. Struwe, *On the Hölder continuity of bounded weak solutions of quasilinear parabolic systems*, Manusc. Math. **35** (1981), 125–145.

[34] M. Struwe, *On partial regularity results for the Navier–Stokes equations*, Comm. on Pure and Appl. Math. **41** (1988), 437–458.

[35] M. Struwe, *Regular Solutions of the Stationary Navier-Stokes equations on* $\mathbb{R}^5$, Math. Ann. **302** (1995), 719–741.

[36] W. von Wahl, *The equations of Navier-Stokes and abstract parabolic equations*, Vieweg, Braunschweig, 1985.

[37] K. O. Widman, *Hölder Continuity of Solutions of Elliptic Systems*, Manuscripta Math. **5** (1971), 299–308.

Jens Frehse and Michael Růžička
Institute of Applied Mathematics
University of Bonn
Beringstraße 4–6
D-53115 Bonn
Germany

KONSTANTIN PILECKAS

Recent Advances in the Theory of Stokes and Navier–Stokes Equations in Domains with Non-compact Boundaries

Abstract

We reflect upon various aspects of the theory of Stokes and Navier-Stokes problems in domains with outlets to infinity.

Contents

1. Introduction

The solvability of the boundary and initial–boundary value problems for Stokes and Navier-Stokes equations is one of the most important questions in the mathematical hydrodynamics. It has been studied in many papers and monographs (e.g. [21], [9]). The existence theory which is developed there concerns mainly the domains with compact boundaries (bounded or exterior). Although some of these results do not depend on the shape of the boundary, many problems of scientific interest concerning the flow of a viscous incompressible fluid in domains with noncompact boundaries were unsolved. This paper is written on the basis of lectures given by the author and is related to questions of flow in domains with noncompact boundaries.

The attention to the question of correct formulation of Stokes and Navier-Stokes problems in domains with noncompact boundaries has been drawn by J. Heywood [11]. In 1976 J. Heywood has shown that in domains with noncompact boundaries the flow of a viscous liquid is not always uniquely determined by applied external forces and by usual initial and boundary conditions. Moreover, certain physically important quantities (as fluxes of the velocity or values of the pressure at infinity) should be prescribed additionally. In [11] the Stokes

$$\begin{aligned} -\nu\Delta u+\nabla p &= f, \qquad \operatorname{div} u=0 \quad \text{in} \quad \Omega \\ u &= 0 \qquad \text{on} \quad \partial\Omega \end{aligned} \tag{1}$$

and the Navier-Stokes problems

$$\begin{aligned} -\nu\Delta u+(u\cdot\nabla)u+\nabla p &= f, \qquad \operatorname{div} u=0 \quad \text{in} \quad \Omega \\ u &= 0 \qquad \text{on} \quad \partial\Omega \end{aligned} \tag{2}$$

were considered in the aperture domain

$$\Omega=\{x=(x_1,x_2,x_3)\in I\!R^3 : x_3\neq 0 \text{ or } x_3=0,\ x'=(x_1,x_2)\in S\}, \tag{3}$$

where S is a bounded region in the plane $I\!R^2$. For such a domain the Stokes problem (1) admits infinitely many solutions with a finite Dirichlet integral[1] and the solution can be specified by prescribing either the total flux of the liquid through the aperture S or the "pressure drop", i.e. the difference between the limits of the pressure $p(x)$ as $|x|\to\infty$, $x_3>0$ and $|x|\to\infty$, $x_3<0$. Analogous problems for the nonlinear stationary Navier-Stokes system were solved in [11] for small data.

After 1976 the theory of Stokes and Navier-Stokes equations in domains with noncompact boundaries has got a great progress in a row of papers (see [22]-[25], [58], [42], [43], [44], [19], [33], [18], [53]-[56]). Problems (1) and (2) were studied in a large class of domains having outlets to infinity, i.e. in domains $\Omega\subset I\!R^n$ which outside

[1] This is $\|\nabla u; L^2(\Omega)\| = \left(\int_\Omega \sum_{i=1}^n \left|\frac{\partial u}{\partial x_i}\right|^2 dx\right)^{1/2}$.

the sphere $|x| = R_0$ split in M connected unbounded components Ω_i: "the outlets to infinity". Problems (1) and (2) were solved in these papers under an additional flux condition. This means that for each outlet Ω_i there is a prescribed number F_i (which is called "the flux through the outlet Ω_i") such that

$$F_i = \int_{\sigma_i(t)} u \cdot n \, ds, \quad \sum_{i=1}^{M} F_i = 0, \tag{4}$$

where $\sigma_i(t) = \{x \in \Omega_i : |x| = t\}$. In particular, the solvability of the nonlinear problem (2), (4) in the aperture domain (3) was proved (see [23]) for arbitrary large data. Another characteristic feature of such problems is the fact that the solution to (1), (4) and (2), (4) often has an infinite Dirichlet integral (this depends on the geometry of the outlets) and the usual methods of energy estimates may become insufficient. O.A. Ladyzhenskaya and V.A. Solonnikov [24] have developed a special technique of integral estimates (so-called "techniques of Saint-Venant's principle") which makes it possible to construct and investigate the solutions with unbounded Dirichlet integral.

In the recent years there was also obtained a progress in the investigation of differential and decay properties and of the asymptotics of the solutions (see [7], [8], [31], [10], [48], [49], [50], [36]).

In the paper we give an update review of known results on these problems and indicate some important open problems. We study Stokes and Navier-Stokes problems in weighted function spaces which are chosen to reflect the decay properties of the solutions. The paper consists of five sections. In Section 2 we present the results concerning the weak solvability of (1), (4) and (2), (4). We start with the very general approach of V.A. Solonnikov [55]. As examples we present the results in domains with outlets having specified geometry which had been obtained in preceding papers (see [11], [23], [24], [58], [43]).

In Section 3 the strong solutions to (1), (4) and (2), (4) are studied in weighted function spaces assuming a "parabolic-like" structure of the outlets. Moreover, the precise asymptotics of the solutions are constructed. The results of this section are based on the papers [48], [49], [50], [36].

Section 4 is devoted to the investigation of strong solutions to (1), (4) and (2), (4) in the aperture domain (3). Mainly, the results are taken from [7], [8].

Finally, in Section 5 we discuss the possibility to extend the obtained results for other types of outlets to infinity (e.g. "layer-like" outlets) and for more complicated equations (e.g. compressible Navier-Stokes equations, noncompact free boundary problems, equations of nonnewtonian fluid motion). Moreover, we outline the possibility of new approaches, consisting of prescribing instead of fluxes and pressure drops much more general asymptotic conditions at infinity, which give better models of real physical phenomena.

The author brings his cordial gratitude to the organizers of the Winter school in Paseky, above all to Professor J. Málek, and to all lecturers and participants of the school. The author is also deeply grateful to Dr. G. Thäter for her help preparing the manuscript.

2. Spaces of divergence free vector fields and weak solutions to Stokes and Navier-Stokes systems

2.1. Basic notations and function spaces

Let Ω be an arbitrary (bounded or unbounded) domain in $\mathbb{R}^n$, $n = 2, 3$, with the boundary $\partial\Omega$. Denote by $C_0^\infty(\Omega)$ the set of all infinitely often differentiable real functions or vector fields with compact supports contained in Ω and by $J_0^\infty(\Omega)$ the subset of all divergence free (i.e. satisfying the condition $\operatorname{div} u(x) = \sum_{k=1}^n \partial u_k(x)/\partial x_k = 0$) vector fields from $C_0^\infty(\Omega)$.

Let $L^q(\Omega)$, $q > 1$, be the set of all Lebesgue integrable functions defined in Ω and having the finite norm

$$\|u; L^q(\Omega)\| = \left(\int_\Omega |u|^q \, dx \right)^{1/q}.$$

We use the same notation for the space of q integrable vector fields in Ω.

$D^q(\Omega)$ and $W^{1,q}(\Omega)$ are the spaces of functions equipped with the norms

$$\|u; D^q(\Omega)\| = \|\nabla u; L^q(\Omega)\| = \left(\int_\Omega \sum_{i=1}^n \left| \frac{\partial u}{\partial x_i} \right|^q dx \right)^{1/q} \tag{5}$$

and

$$\|u; W^{1,q}(\Omega)\| = \left(\int_\Omega \left(|u|^q + \sum_{i=1}^n \left| \frac{\partial u}{\partial x_i} \right|^q \right) dx \right)^{1/q}. \tag{6}$$

$D_0^q(\Omega)$ and $W_0^{1,q}(\Omega)$ are the closures of $C_0^\infty(\Omega)$ in the norms (5) and (6), respectively. $\widehat{H}^q(\Omega)$ and $\widehat{J}_0^{1,q}(\Omega)$ are the subspaces of divergence free vectors in $D_0^q(\Omega)$ and $W_0^{1,q}(\Omega)$. $H^q(\Omega)$ and $J_0^{1,q}(\Omega)$ are the closures of $J_0^\infty(\Omega)$ in the norms (5) and (6). For $q = 2$ we put $D^2(\Omega) = D(\Omega)$, $D_0^2(\Omega) = D_0(\Omega)$, etc. It is obvious that

$$\widehat{H}^q(\Omega) \supset H^q(\Omega), \quad \widehat{J}_0^{1,q}(\Omega) \supset J_0^{1,q}(\Omega).$$

Furthermore, there holds the following result:

Theorem 1 *Let $\Omega \subset \mathbb{R}^n$ be a bounded or exterior domain with Lipschitz boundary $\partial\Omega$. Then*

$$\widehat{H}^q(\Omega) = H^q(\Omega), \quad \widehat{J}_0^{1,q}(\Omega) = J_0^{1,q}(\Omega) \quad \forall\, q > 1.$$

The proof of Theorem 1 for $q = 2$ can be found in [22] and for arbitrary $q > 1$ in [44].

2.2. General domains with outlets to infinity

Let us introduce domains having outlets to infinity. Following V.A. Solonnikov [55], we consider the most general case first and then discuss concrete examples.

Let us consider an unbounded domain $\Omega \subset \mathbb{R}^n$, $n = 2, 3$, having M outlets to infinity. We suppose that outside the sphere $|x| = R_0$ the domain Ω splits into M unbounded connected components Ω_i (the outlets to infinity). Suppose that for each $i = 1, \ldots, M$ there exist bounded subsets Ω_{ik}, $k = 1, 2, \ldots$, of Ω_i exhausting Ω_i as $k \to \infty$. We put

$$\Omega_{(k)} = \Omega_0 \cup \Omega_{1k} \cup \ldots \cup \Omega_{Mk}, \quad \omega_{ik} = \Omega_{ik+1} \backslash \Omega_{ik}$$

and assume the following conditions:

(I) $\Omega_{ik} \subset \Omega_{ik+1}$;

(II) $\operatorname{dist}(\Omega_{ik}, \Omega_i \backslash \Omega_{ik+1}) > 0, \qquad \operatorname{dist}(\Omega_0, \Omega_i \backslash \Omega_{ik}) \to \infty \qquad$ *as* $k \to \infty$;

(III) $\Omega_{(k)}$ *are connected Lipschitz domains;*

(IV) *For an arbitrary divergence free vector field* $u \in D^q(\Omega_{ik+1})$ *vanishing at* $\partial\Omega \cap \partial\Omega_{ik+1}$ *and having a zero flux through* Ω_i, *i.e.*

$$\int_{\sigma_i} u \cdot n\, ds = 0, \quad \sigma_i = \Omega_i \cap \partial B(0, R_0), \quad B(0, R_0) = \{x : \ |x| < R_0\},$$

there exists a divergence free vector field $U_{ik} = P_{ik}u \in D^q(\Omega_i)$ *such that*

$$\begin{aligned} U_{ik} &= u \quad \text{in} \quad \Omega_{ik}, \quad U_{ik} = 0 \quad \text{in} \quad \Omega_i \backslash \Omega_{ik+1}, \\ U_{ik} &= 0 \quad \text{on} \quad \partial\Omega_i \cap \partial\Omega. \end{aligned}$$

Let us shortly discuss condition (**IV**). The vector field U_{ik} can be constructed as the sum

$$U_{ik} = \chi_{ik} u + \widehat{U}_{ik},$$

where χ_{ik} are smooth cut-off functions, $\chi_{ik}(x) = 1$, $x \in \Omega_{ik}$, $\chi_{ik}(x) = 0$, $x \in \Omega_i \backslash \Omega_{ik+1}$ and $0 \le \chi_{ik}(x) \le 1$, and $\widehat{U}_{ik}$ is the solution of the problem

$$\operatorname{div} \widehat{U}_{ik} = -u \cdot \nabla \chi_{ik} \quad \text{in} \quad \omega_{ik}, \quad \widehat{U}_{ik} = 0 \quad \text{on} \quad \partial\omega_{ik}. \tag{7}$$

($\widehat{U}_{ik}$ are extended by zero in $\Omega \backslash \omega_{ik}$.) Thus, the construction of $U_{ik} = P_{ik}u$ is based on the solvability of the divergence equation

$$\operatorname{div} w = g \quad \text{in } G, \quad w = 0 \quad \text{on } \partial G, \tag{8}$$

where $G \subset \mathbb{R}^n$ is a bounded domain. There holds the following

Theorem 2 *Let $G \subset I\!R^n$ be a bounded domain with Lipschitz boundary and let $g \in \widehat{L}^q(G) = \{\varphi \in L^q(G) : \int_G \varphi(x)\, dx = 0\}$. Then the problem (8) has a solution $w \in D_0^q(G)$ satisfying the estimate*

$$\|w; D^q(G)\| \leq c\, \|g; L^q(G)\|. \tag{9}$$

The solvability of divergence equation (8) in Lipschitz domains was first proved for $q = 2$ by O.A. Ladyzhenskaya and V.A. Solonnikov (1976) [22]. The result of Theorem 2 for arbitrary $q > 1$ was obtained by K. Pileckas (1980,1983) [42], [44] and independently by M.E. Bogovskii (1980) [6].

For domains with outlets to infinity the spaces $\widehat{H}^q(\Omega)$ and $H^q(\Omega)$ are not necessarily identical. There holds

Theorem 3 (i) *Suppose that the domain Ω satisfies the conditions* (**I**)–(**IV**) *and that*

$$\lim_{k\to\infty} \|P_{ik}u; D^q(\omega_{ik})\| = 0, \quad i = 1, \dots, M, \tag{10}$$

for arbitrary divergence free $u \in D^q(\Omega_i)$ such that

$$u = 0 \quad on\ \partial\Omega \cap \partial\Omega_i, \quad \int_{\sigma_i} u \cdot n\, ds = 0. \tag{11}$$

Then the space $H^q(\Omega)$ consists of all $u \in \widehat{H}^q(\Omega)$ satisfying the condition

$$\int_{\sigma_i} u \cdot n\, ds = 0, \quad i = 1, \dots, M. \tag{12}$$

(ii) *Suppose, in addition, that there exist divergence free vector fields $W_i \in D^q(\Omega_i)$, $i = 1, \dots, s$, $s \leq M$, with*

$$W_i = 0 \quad on\ \partial\Omega \cap \partial\Omega_i, \quad \int_{\sigma_i} W_i \cdot n\, ds = 1 \tag{13}$$

and that there are no such fields in $\Omega_{s+1}, \dots, \Omega_M$, i.e. $\int_{\sigma_i} u \cdot n\, ds = 0$, $i = s+1, \dots, M$, $\forall\, u \in \widehat{H}^q(\Omega)$. Then

$$\dim \widehat{H}^q(\Omega)/H^q(\Omega) = s - 1.$$

The first part of this theorem can be proved by approximating arbitrary $u \in \widehat{H}^q(\Omega)$, satisfying (12), by a sequence of divergence free fields from $J_0^\infty(\Omega)$. To do so, one should first construct $\{u_k\}$, $u_k \in H^q(\Omega)$, $\mathrm{supp} u_k \subset \Omega_{(k+1)}$ such that $\|u - u_k; D^q(\Omega)\| \to 0$ for $k \to \infty$. At this place the conditions (**IV**) and (10) are used. Now, since the spaces $H^q(G)$ and $\widehat{H}^q(G)$ coincide for bounded Lipschitz domains G (see Theorem 1), each u_k may be approximated by vector fields from $J_0^\infty(\Omega)$.

The second part is proved by constructing vectors $A_j \in \widehat{H}^q(\Omega)$ for $j = 1, \dots, s-1$, satisfying the condition

$$\int_{\sigma_j} A_j \cdot n\, ds = 1, \quad \int_{\sigma_s} A_j \cdot n\, ds = -1, \quad \int_{\sigma_k} A_j \cdot n\, ds = 0, \quad k \neq j, k \neq s.$$

These vectors form a basis in $\widehat{H}^q(\Omega)/H^q(\Omega)$.

Theorem 3 has been proved by V.A. Solonnikov [55]. In [55] one can also find the construction of divergence free vector fields which create fluxes through the outlets Ω_i for general domains satisfying **(I)**–**(IV)**. Below we discuss examples of such domains and we give hints how to construct these vector fields in concrete cases. Another approach how to construct the vector fields satisfying the flux condition (4) is presented in [19].

Remark. Analogous results are also obtained for the spaces $\widehat{J}_0^{1,q}(\Omega)$ and $J_0^{1,q}(\Omega)$. In this case one should assume the existence of divergence free vector fields belonging to $W^{1,q}(\Omega_i)$ and having nonzero fluxes through the outlets to infinity. Let us mention that for $q = 2$ such divergence free vector fields do not exist for a large class of two-dimensional domains with outlets to infinity. This result has been proved by L.V. Kapitanskii [17] for the domain $\Omega \subset I\!R^2$ such that for every outlet Ω_i the boundary $\partial\Omega_i$ has only one noncompact connected component Γ_i and Γ_i is a Jordan curve dividing the entire plane into two parts. The boundary $\partial\Omega_i$ of the domain Ω may have an arbitrary number of compact components, bounded by "holes" and there are no assumptions on the smoothness of the compact components of $\partial\Omega$. Thus, for the domains with the described properties

$$\widehat{J}_0^{1,2}(\Omega) = J_0^{1,2}(\Omega).$$

2.3. Spaces of functions with an infinite Dirichlet integral

Let, in addition to **(I)**–**(IV)**, the domain Ω satisfy the condition

(V) *The operator P_{ik} is uniformly bounded, i.e.*

$$\|P_{ik}u; D^q(\omega_{ik})\| \leq b_1 \|u; D^q(\omega_{ik})\|, \tag{14}$$

where b_1 is independent of k.

Let r_k be positive numbers satisfying the conditions:

$$r_{k+1} \geq r_k, \quad r_{k+l} \leq a_0 a_1^l r_k, \quad \forall\, k, l > 0, \tag{15}$$

with $a_0 > 0$, $a_1 > 0$.

$D(\Omega; r)$, $r = (r_1, \dots, r_k, \dots)$, is the space of vector fields defined in Ω, equal to zero on $\partial\Omega$ and having the finite norm

$$\|u; D(\Omega; r)\|^2 = \sup_{k \geq 1} (r_k^{-1} \|u; D(\Omega_{(k)})\|^2). \tag{16}$$

$H(\Omega; r)$ is the subspace of divergence free vector fields from $D(\Omega; r)$ with zero fluxes through all the outlets Ω_i.

$H^*(\Omega; r)$ denotes the space of vector valued distributions with the norm

$$\|f; H^*(\Omega; r)\|^2 = \sup_{k\geq 1}(r_k^{-1}\|f; H^*(\Omega_{(k)})\|^2),$$

where $H^*(G)$ is the dual space to $H(G)$.

The spaces $D(\Omega; r)$ describe the growth of the Dirichlet integral over the domains $\Omega_{(k)}$ as $k \to \infty$. To describe this growth over domains ω_{ik}, let us introduce another family of spaces. Let κ_0 and κ_{ik} be positive numbers such that

$$\begin{aligned} \kappa_{ik+l} &\leq a_0 a_1^l \kappa_{ik}, \quad \forall\, k, l > 0, \\ \kappa_{ik-l} &\leq a_0 a_1^l \kappa_{ik}, \quad \forall\, k > 0,\ l = 1, \ldots, k-1. \end{aligned} \tag{17}$$

$\mathcal{D}(\Omega; \kappa)$, $\kappa = (\kappa_0, \kappa_{11}, \kappa_{12}, \ldots, \kappa_{21}, \ldots)$, is the space of vector fields equal to zero on $\partial\Omega$ with the finite norm

$$\|u; \mathcal{D}(\Omega; \kappa)\|^2 = \max\left(\kappa_0^{-1}\|u; D(\Omega_{(1)})\|^2, \quad \sup_{i,k>0}(\kappa_{ik}^{-1}\|u; D(\omega_{ik})\|^2\right). \tag{18}$$

$\mathcal{H}(\Omega; \kappa)$ is the subspace of divergence free vector fields from $\mathcal{D}(\Omega; \kappa)$ with zero fluxes through all outlets to infinity and $\mathcal{H}^*(\Omega; \kappa)$ is the space of vector valued distributions with

$$\|f; \mathcal{H}^*(\Omega; \kappa)\|^2 = \max(\kappa_0^{-1}\|f; H^*(\Omega_{(2)})\|^2, \sup_{i,k>0}(\kappa_{ik}^{-1}\|f : H^*(\omega_{ik} \cup \omega_{ik+1})\|^2) < \infty.$$

Similar norms can be defined for every outlet Ω_i. For example,

$$\|u; D(\Omega_i; r_i)\|^2 = \sup_k(r_{ik}^{-1}\|u; D(\Omega_{ik})\|^2).$$

The above defined spaces were originally introduced and investigated by V.A. Solonnikov [55]. Here we omit their detailed properties and mention that because of (15), (17) the change Ω_k, ω_{ik} to Ω_{k+l}, ω_{ik+l} in the definitions of norms produces equivalent norms. We also notice that the following inequalities are valid:

$$\begin{aligned} \|u; D(\Omega; r)\| &\leq \|u; \mathcal{D}(\Omega; \kappa)\|, \\ \|u; H^*(\Omega; r)\| &\leq \|u; \mathcal{H}^*(\Omega; \kappa)\|, \end{aligned}$$

where $r_k = \sum_{i=1}^M \sum_{j=1}^{k-1} \kappa_{ij} + \kappa_0$.

2.4. Stokes and Navier-Stokes problems

Let us consider the boundary value problems for the Stokes and Navier-Stokes equations (1) and (2) with the additional flux condition (4). Theorems 4-8 below belong to V.A. Solonnikov [55].

We start by considering the Stokes problem (1), (4). By a weak solution of (1), (4) we mean a divergence free vector field u with a finite Dirichlet integral in every bounded subdomain $\Omega' \subset \Omega$, vanishing on $\partial\Omega$, satisfying the flux condition (4) and the integral identity

$$\nu \int_\Omega \nabla u : \nabla \eta \, dx = \int_\Omega f \cdot \eta \, dx \tag{19}$$

for any $\eta \in H(\Omega)$ with a compact support.

Theorem 4 *Suppose that Ω satisfies the conditions* (**I**)–(**V**) *and that there exists a divergence free vector field $A \in D(\Omega; r)$ such that*

$$\int_{\sigma_i} A \cdot n \, ds = F_i, \quad i = 1, \dots, M. \tag{20}$$

Moreover, assume that the constants a_1 and b_1 in the inequalities (15) *and* (14) *are related by*

$$a_1 < (b_1 + 1) b_1^{-1}. \tag{21}$$

Then the problem (1), (4) *with $f \in H^*(\Omega; r)$ has a unique weak solution $u \in D(\Omega; r)$ and there holds the estimate*

$$\|u; D(\Omega; r)\| \le c\,(\|f; H^*(\Omega; r)\| + \|A; D(\Omega; r)\|). \tag{22}$$

The next theorem concerns estimates of a weak solution in an arbitrary outlet to infinity.

Theorem 5 *Let f and A have bounded norms $\|f; H^*(\Omega_i; r_i)\|$ and $\|A; D(\Omega_i; r_i)\|$, where the numbers r_i satisfy conditions* (15) *and* (21) *and let u be a weak solution of* (1), (4) *such that*

$$\lim_{k\to\infty} \left(\frac{\mu}{1+\mu}\right)^k \|u; D(\Omega_{ik})\|^2 = 0, \quad (1+\mu)/\mu > a_1, \quad \mu > 0.$$

Then

$$\begin{aligned} \|u; D(\Omega_i \backslash \Omega_{i1}; r_i)\| \le \ c \ & (\|f; H^*(\Omega_i \backslash \Omega_{i1}; r_i)\| + \|A; D(\Omega_i \backslash \Omega_{i1}; r_i)\| \\ & + \|u; D(\omega_{i1})\| + \|A; D(\omega_{i1})\|). \end{aligned} \tag{23}$$

In particular, if $f \in H^*(\Omega_i)$, $A \in D(\Omega_i)$ (i.e. $r_{i1} = \ldots = r_{ik} = \ldots = 1$ and A has a finite Dirichlet integral over Ω_i), then $u \in D(\Omega_i)$.

Now we estimate the Dirichlet integral of the weak solution in ω_{ik}.

Theorem 6 *Let f and A have finite norms $\|f;\mathcal{H}^*(\Omega_i\backslash\Omega_{i1};\kappa_i)\|$ and $\|A;\mathcal{D}(\Omega_i\backslash\Omega_{i1};\kappa_i)\|$, where the numbers κ_{ik} satisfy* (17) *and*

$$a_1 < (b_1+2)(b_1+1)^{-1}. \tag{24}$$

Then a weak solution u of the problem (1), (4) *satisfies the estimate*

$$\begin{aligned}\|u;\mathcal{D}(\Omega_i\backslash\Omega_{i1};\kappa_i)\| \leq \ c \ & (\|f;\mathcal{H}^*(\Omega_i\backslash\Omega_{i1};\kappa_i)\| + \|A;\mathcal{D}(\Omega_i\backslash\Omega_{i1};\kappa_i)\| \\ & +\|u;D(\omega_{i1})\| + \|A;D(\omega_{i1})\|).\end{aligned} \tag{25}$$

Finally, we present the result concerning the solvability of the Stokes problem in the spaces $\mathcal{D}(\Omega;\kappa)$.

Theorem 7 *Assume that there exists a divergence free vector field $A\in\mathcal{D}(\Omega;\kappa)$ satisfying the flux condition* (4) *and let $f\in\mathcal{H}^*(\Omega;\kappa)$. If there holds the inequality* (24), *then the problem* (1), (4) *has a unique solution $u\in\mathcal{D}(\Omega;\kappa)$ and*

$$\|u;\mathcal{D}(\Omega;\kappa)\| \leq c\left(\|f;\mathcal{H}^*(\Omega;\kappa)\| + \|A;\mathcal{D}(\Omega;\kappa)\|\right). \tag{26}$$

Let us turn to the nonlinear problem (2), (4). A weak solution of (2), (4) is defined as a divergence free vector field $u\in D(\Omega')$ for arbitrary bounded $\Omega'\subset\Omega$, vanishing on $\partial\Omega$, satisfying the condition (4) and the integral identity

$$\nu\int_\Omega \nabla u : \nabla\eta\, dx + \int_\Omega (u\cdot\nabla)u\cdot\eta\, dx = \int_\Omega f\cdot\eta\, dx$$

with arbitrary $\eta\in H(\Omega)$, having a compact support. There holds the following existence result.

Theorem 8 *Let the domain Ω satisfy the conditions* (**I**)–(**V**) *and in addition:*

(1) *for each $v\in D(\omega_{ik})$ with $v=0$ on $\partial\Omega\cap\partial\omega_{ik}$ the inequality*

$$\|v;L^4(\omega_{ik})\| \leq \gamma_{ik}\|v;D(\omega_{ik})\| \tag{27}$$

is valid;

(2) *there exists a divergence free vector field $A\in D(\Omega;r)$, vanishing on $\partial\Omega$ and satisfying the flux condition* (4), *such that*

$$\int_{\Omega_{(k)}} |A|^2|v|^2\, dx \leq \delta\int_{\Omega_{(k)}} |\nabla v|^2\, dx \tag{28}$$

holds for arbitrary $v\in D(\Omega_{(k)})$, $v=0$ on $\partial\Omega\cap\partial\Omega_{(k)}$ with

$$\delta < \frac{\nu}{2(1+b_1^2)^{1/2}};$$

(3) *the numbers r_k satisfy the inequality*

$$r_k \geq c_*(r_{k+1} - r_k) + c^*\gamma_k^2(r_{k+1} - r_k)^{3/2}, \qquad \text{where } \gamma_k^2 = \sum_{i=1}^{M} \gamma_{ik}^2. \tag{29}$$

Then the problem (2), (4) *with arbitrary* $f \in H^*(\Omega; r)$ *has at least one weak solution.*

As in the linear case, it is possible to prove that if the vector field A has a finite Dirichlet integral over some outlet Ω_i, then the problem (2), (4) has a weak solution with the finite norm $\|u; D(\Omega_i)\|$.

Note, that a vector field $v \in D(\Omega; r)$ with r_k satisfying (29) may have an infinite Dirichlet integral in Ω, if there holds the relation

$$\sum_{k=1}^{\infty} \gamma_k^{-4/3} = \infty. \tag{30}$$

Under certain additional conditions it is possible to prove that for sufficiently small data the weak solution of the Navier-Stokes problem (2), (4) is unique in the class of functions, satisfying the relation

$$\lim_{k\to\infty} \left(\frac{\mu_*}{1+\mu_*}\right)^k \|v; D(\Omega_{(k)})\|^2 = 0,$$

where $\mu_* > 0$ is a constant depending on Ω.

The formulated existence and uniqueness results were proved in V.A. Solonnikov [55] by applying difference inequalities techniques which are very close to the techniques of differential inequalities for energy integrals ("techniques of Saint Venant's principle") developed by O.A. Ladyzhenskaya and V.A. Solonnikov [24], [25] for concrete outlets to infinity (in this context see also the papers [15], [16], [60]).

2.5. Examples

Example 1. Paraboloid-like outlets to infinity. Let $\Omega \subset I\!\!R^n$, $n = 2, 3$, be a domain with outlets to infinity which have the following form

$$x^{i\prime} g_i^{-1}(x_n^i) \in \sigma_i, \quad x_n^i > 0, \tag{31}$$

where $\{x^i\}$ are appropriately chosen systems of coordinates and $x^{i\prime} = (x_1^i, \ldots, x_{n-1}^i)$. Below we omit the index i in notations for local coordinates. Thereby σ_i is a bounded domain in $I\!\!R^{n-1}$ with Lipschitz boundary. We assume that the function g_i satisfies the conditions

$$|g_i(t) - g_i(t')| \leq L_i\,|t - t'|, \quad \forall\, t, t' > 0, \quad g_i(t) \geq g_0 > 0, \tag{32}$$

$$\lim_{t\to\infty} \tfrac{d}{dt} g_i(t) = 0, \quad |\tfrac{d}{dt} g_i(t)| \le L_i, \quad \forall\, t > 0. \tag{33}$$

We introduce the following notations

$$\begin{aligned}
\sigma_i(t) &= \{x \in \Omega_i : x_n = t\}, \quad \text{i.e. } \sigma_i = \sigma_i(0), \\
R_{i0} &= 0, \quad R_{ik+1} = R_{ik} + (2L_i)^{-1} g_{ik}, \quad g_{ik} = g_i(R_{ik}), \\
\Omega_{ik} &= \{x \in \Omega_i : x_n < R_{ik}\}, \quad \omega_{ik} = \Omega_{ik+1} \backslash \Omega_{ik}, \quad i = 1, \ldots, M.
\end{aligned}$$

It is not difficult to show:

Lemma 9 *There hold the relations*

$$\frac{1}{2}\, g_{ik} \le g_i(t) \le \frac{3}{2}\, g_{ik}, \quad t \in [R_{ik}, R_{ik+1}], \tag{34}$$

$$\int_0^\infty g_i^{-1}(\tau)\, d\tau = \infty. \tag{35}$$

Using the condition (34) it is shown in [58] that the constant in the estimate (9) for the solution of the divergence equation (8) in the domain ω_{ik} may be chosen independently of k. Let χ_{ik} be the cut-off function defined after condition **(IV)**. It is easy to understand that χ_{ik} can be taken to satisfy the inequalities

$$|\nabla \chi_{ik}(x)| \le c\, g_{ik}^{-1}.$$

Then the vector field $\widehat{U}_{ik}$ defined in (7) admits the estimate

$$\|\widehat{U}_{ik}; D^q(\omega_{ik})\| \le c\, g_{ik}^{-1} \|u; L^q(\omega_{ik})\|.$$

Hence, the vector $U_{ik} = \chi_{ik} u + \widehat{U}_{ik}$ satisfies the inequality

$$\|U_{ik}; D^q(\omega_{ik})\| \le \|u; D^q(\omega_{ik})\| + c\, g_{ik}^{-1} \|u; L^q(\omega_{ik})\|.$$

Since by Friedrich's inequality

$$\|u; L^q(\omega_{ik})\| \le c\, g_{ik} \|u; D^q(\omega_{ik})\|$$

with c independent of k, we get that

$$\|U_{ik}; D^q(\omega_{ik})\| \le c\, \|u; D^q(\omega_{ik})\|.$$

If $u \in D^q(\Omega_i)$, the right-hand side of the last inequality tends to zero and, hence we have checked the condition (10) for the operator P_{ik}.

It has been proved in [44] that every divergence free vector field $u \in D^q(\Omega_i)$ vanishing on $\partial\sigma_i$ has a zero flux through $\sigma_i(t)$, provided that

$$\int_0^\infty g_i^{(1-n)(q-1)-q}(t)\, dt = \infty. \tag{36}$$

This follows from the estimates

$$\begin{aligned}|F_i|^q &= \left|\int_{\sigma_i(t)} u\cdot n\,ds\right|^q \le c\,g_i^{(n-1)(q-1)}(t)\int_{\sigma_i(t)}|u|^q\,dx' \\ &\le c\,g_i^{(n-1)(q-1)+q}(t)\int_{\sigma_i(t)}|\nabla u|^q\,dx'\,.\end{aligned}$$

Since the flux F_i does not depend on t it follows that

$$|F_i|^q\int_0^R g^{(1-n)(q-1)-q}(t)\,dt \le c\int_{\Omega_i}|\nabla u|^q\,dx < \infty.$$

Passing $R\to\infty$ one can see that under (36) the last inequality is possible only if $F_i = 0$.

Let us construct now the divergence free vector fields satisfying the flux condition (4). Denote by γ_{ij} the contour consisting of two lines γ_i and γ_j lying in the domains Ω_i and Ω_j, respectively, and of a smooth curve $\tilde{\gamma}_{ij}$ which joins them being situated in Ω so that the distance from γ_{ij} to the surface $\partial\Omega$ is not less than a number $d_0 > 0$. First we consider the case $n = 3$. We introduce the divergence free vector fields

$$A^{ij}(x) = \operatorname{curl}\left(\zeta^{ij}(x)\cdot b^{ij}(x)\right) = \nabla\zeta^{ij}(x)\times b^{ij}(x),$$

where

$$b^{ij}(x) = \oint_{\gamma_{ij}}\frac{x-y}{|x-y|^3}\times dl \quad\text{and}\quad \zeta^{ij}(x) = \psi\left(\ln\frac{\rho(\delta^{ij}(x))}{\Delta(x)}\right).$$

Here $\psi(t)$ and $\rho(t)$ are infinitely often differentiable, monontone functions with $\psi(t) = 0$ for $t\le 0$, $\psi(t) = 1$ for $t\ge 1$, $\rho(t) = a_1d_0/2$ for $t\le a_2d_0/2$, $\rho(t) = t$ for $t\ge a_2d_0$, a_1 and a_2 are positive constants, $\delta^{ij}(x)$ and $\Delta(x)$ are the regularized distances from the point x to γ_{ij} and $\partial\Omega$ respectively. The regularized distance from x to the closed set $G\subset I\!R^n$ is an infinitely often differentiable in $I\!R^n\backslash G$ function, satisfying the inequalities

$$c_1d_G(x)\le\Delta_G(x)\le c_2d_G(x),\quad |D^\alpha\Delta_G(x)|\le c_3d_G^{1-|\alpha|}(x),$$

where $d_G(x)$ is the real distance from x to G (see [59]).

The divergence free vector fields $A^{ij}(x)$ are infinitely often differentiable in Ω and vanish near $\partial\Omega$ and near the contour γ_{ij}. Moreover,

$$\int_{\sigma_k}A^{ij}\cdot n\,ds = \pm 4\pi(\delta_{ik}+\delta_{jk}).$$

Here one has the sign " - ", if $k = i$ or $k = j$ and if the integration over the part γ_k of the contour γ_{ij} is in the direction of increase of the coordinate x_3, and the vector $n(x)$ points in the same direction. Finally, A^{ij} satisfy the estimates

$$|D^\alpha A^{ij}(x)|\le\frac{C}{g_i^{2+|\alpha|}(x_3)},\quad x\in\Omega_i,\quad |\alpha|\ge 0. \tag{37}$$

From (37) it follows that $A^{ij}(x)$ can only have a finite Dirichlet integral over Ω_i if

$$\int_0^\infty \frac{dt}{g_i^{(n-1)(q-1)+q}(t)} < \infty, \quad n = 3.$$

The analogous results are valid also for the two-dimensional case. The role of the vectors A^{ij} is played now by

$$A^{ij}(x) = (\zeta^{ij}_{x_1}, \ -\zeta^{ij}_{x_2}),$$

where the function $\zeta^{ij}(x)$ is defined by the same formula as for $n = 3$. Then instead of (37) there holds the estimate

$$|D^\alpha A^{ij}(x)| \leq \frac{C}{g_i^{1+|\alpha|}(x_3)}, \quad x \in \Omega_i, \ |\alpha| \geq 0. \tag{38}$$

The vector field A satisfying the flux condition (4) can be taken as a linear combination of $A^{i,i+1}$, i.e.

$$A(x) = \sum_{i=1}^{M} \alpha_i A^{i,i+1}(x),$$

where the constants α_i are chosen to satisfy the condition (4).
By Theorem 3 we conclude that

$$\dim \widehat{H}^q(\Omega)/H^q(\Omega) = s - 1, \quad s \leq M,$$

where s is a number of the outlets to infinity for which the integrals

$$\int_0^\infty \frac{dt}{g_i^{(n-1)(q-1)+q}(t)} \tag{39}$$

are finite.

The similar theory can be developed for more general weighted spaces of divergence free vector fields. Let $H^q_{æ}(\Omega), æ = (æ_1, \ldots, æ_M), q = (q_0, q_1, \ldots, q_M)$, be the closure of $J_0^\infty(\Omega)$ in the norm

$$\|u; H^q_{æ}(\Omega)\| = \left(\int_{\Omega_{R_0}} |\nabla u|^{q_0}\, dx\right)^{1/q_0} + \sum_{i=1}^{M} \left(\int_{\Omega_i} g_i^{q_i æ_i}(x_3)|\nabla u|^{q_i}\, dx\right)^{1/q_i} \tag{40}$$

and let $\widehat{H}^q_{æ}(\Omega)$ be the space of all divergence free vector functions, having zero traces on $\partial\Omega$ and finite norm (40). If $q_0 = \ldots = q_M$ and $æ_1 = \ldots = æ_M = 0$, we have $H^q_0(\Omega) = H^q(\Omega)$, $\widehat{H}^q_0(\Omega) = \widehat{H}^q(\Omega)$.

There holds the following

Theorem 10 *If among the integrals*

$$\int_0^\infty g_i^{-(n-1)(q_i-1)-q_i+q_i æ_i}(t)\,dt, \qquad i=1,\ldots,M,$$

there are exactly s which converge, then

$$\dim \widehat{H}^q_æ(\Omega)/H^q_æ(\Omega) = s-1.$$

The space $H^q_æ(\Omega)$ consists of those and only those $u \in \widehat{H}^q_æ(\Omega)$ which have zero fluxes through all outlets to infinity.

Let us discuss now the conditions for the solvability of the Stokes problem (1), (4) in the domain Ω with outlets to infinity of the form (31). The numbers κ_{ik} in the definitions for norms of vector fields with an infinite Dirichlet integral can be taken as follows

$$\kappa_{ik} = G^{2æ_i}(x)\exp(-2\beta_i\alpha_i(x_n)), \tag{41}$$

where $G(x) = g_i(x)g_{ik_0}^{-1}$, $\alpha_i(t) = \int_0^t g_i^{-1}(s)\,ds$. The conditions (17), (21), (24) are satisfied for $k > k_0$ if we assume k_0 to be sufficiently large and β_i to be sufficiently small (i.e. $|\beta_i| < \beta_*$). In order to check (17), (21), (24) one should use the relations (32)-(35) and the inequality

$$\mu_*|l-k| \le |\alpha_{il}-\alpha_{ik}| \le \mu^*|l-k|, \quad \alpha_{ik} = \alpha(R_{ik})$$

(μ_*, μ^* are independent of l and k) which follows from (34) (the details can be found in [48]). Finally, we derive the following results.

Theorem 11 (i) *Let $f \in \mathcal{H}^*(\Omega,\kappa)$ with κ defined by (41) and $|\beta_i| < \beta_*$. Then the problem (1), (4) with zero fluxes (i.e. $F_1 = \ldots = F_M = 0$) has a unique weak solution $u \in \mathcal{D}(\Omega;\kappa)$ satisfying the estimate*

$$\|u;\mathcal{D}(\Omega;\kappa)\| \le c\,\|f;\mathcal{H}^*(\Omega;\kappa\|. \tag{42}$$

In particular,

$$\int_{\omega_{ik}} |\nabla u|^2\,dx \le c g_{ik}^{-2æ_i}\exp(-2\beta_i\alpha_{ik})\|f;\mathcal{H}^*(\Omega;\kappa)\|^2. \tag{43}$$

(ii) *Let the integrals (39) with $q=2$ be finite for $i=1,\ldots,s$ and infinite for $i = s+1,\ldots,M$. Suppose that $f \in H^*(\Omega)$ and that $F_i = 0$ for $i = s+1,\ldots,M$. Then there exists a unique weak solution u of problem (1), (4) such that $u \in \widehat{H}(\Omega)$ and identity (19) is valid for each $\eta \in H(\Omega)$. There holds the estimate*

$$\|u;D(\Omega)\|^2 \le c\Big(\|f;H^*(\Omega)\|^2 + \sum_{i=1}^r |F_i|^2\Big). \tag{44}$$

(iii) *Let $f = 0$. Then for arbitrary fluxes $F_i \in I\!R^1, i = 1, \dots, M$, there exists a unique weak solution u to (1), (4) with an infinite Dirichlet integral. The following estimate holds true*

$$\int_{\Omega_{(k)}} |\nabla u|^2\, dx \le c \left(\sum_{i=1}^{M} |F_i|^2 \right) \sum_{k=1}^{M} \int_0^{R_{i\,k+1}} g_i^{-4}(t)\, dt. \tag{45}$$

Concerning the nonlinear Navier–Stokes problem (2), (4) we have the following result:

Theorem 12 ([24], [55]) *Let $f = 0$. Suppose that in addition to (32), (33) the functions g_i satisfy the conditions*

$$\int_0^\infty g_i^{-4/3}(t)\, dt = \infty, \quad i = 1, \dots, M, \tag{46}$$

$$|g_i'(t) g_i^{1/3}(t)| \le \gamma << 1 \quad for\ \ t > k_0, \quad i = 1, \dots, M. \tag{47}$$

Then for arbitrary fluxes $F_i \in I\!R^1, i = 1, \dots, M$, there exists at least one weak solution u to (2), (4) with an infinite Dirichlet integral. This solution admits the estimates

$$\int_{\Omega_{(k)}} |\nabla u|^2\, dx \ \le\ c(|F|) \sum_{k=1}^{M} \int_0^{R_{ik+1}} g_i^{-4}(t)\, dt, \tag{48}$$

$$\int_{\omega_{ik}} |\nabla u|^2\, dx \ \le\ c(|F|) g_{ik}^{-8/3}, \quad k > k_0, \quad i = 1, \dots, M, \tag{49}$$

where $|F| = \left(\sum_{k=1}^{M} |F_i|^2\right)^{1/2}$. Moreover, if the fluxes $F_i, i = 1, \dots, M$, are sufficiently small, then any other (different from u) solution u' satisfies the relation

$$\liminf_{k \to \infty} k^{-3} \int_{\Omega_{(k)}} |\nabla (u' - u)|^2\, dx > 0. \tag{50}$$

In the outlets to infinity Ω_i where the integrals

$$\int_0^\infty g_i^{-4}(t)\, dt \tag{51}$$

are finite it is possible to prescribe instead of fluxes F_i the limiting values for the pressure function p. Such problem was studied by V.A. Solonnikov [53]. Let us present the corresponding results for the more general nonlinear Navier-Stokes problem (2). We assume that the integrals (51) are finite for $i = 1, \dots, s$ and infinite for $i = s + 1, \dots, M$. Consider in Ω the problem (2) with the additional conditions

$$p_j - p_s = p_{*j}, \quad j = 1, \dots, s - 1, \tag{52}$$

where

$$p_i = \lim_{x \in \Omega_i,\ |x| \to \infty} p(x), \quad i = 1, \dots, s. \tag{53}$$

A weak solution to problem (2), (52) is defined as a vector field $u \in \widehat{H}(\Omega)$ satisfying the integral identity

$$\nu \int_\Omega \nabla u : \nabla \eta \, dx - \int_\Omega u\,(u \cdot \nabla)\eta \, dx = \int_\Omega f \cdot \eta \, dx - \sum_{i=1}^{s-1} p_{*j} \int_{\sigma_i} \eta \cdot n \, ds \tag{54}$$

for every η which can be represented as a sum

$$\eta = \sum_{i=1}^{s-1} \lambda_i A^{ii+1} + \varphi, \quad \varphi \in J_0^\infty(\Omega), \ \ \lambda_i \in I\!R^1.$$

Theorem 13 ([53]) *Assume that for all divergence free* $\varphi \in D(\Omega_{(k)})$ *vanishing on* $\partial\Omega_{(k)} \bigcup_{i=1}^{s} \sigma_i(R_k)$ *there holds the estimate*

$$\Big| \int_{\Omega_{(k)}} f \cdot \varphi \, dx \Big| \leq C_f \|\varphi;\ D(\Omega_{(k)})\|$$

with C_f *independent of* k *and* φ*. Then problem* (2), (52) *has at least one weak solution.*

If, in addition, $\int_0^\infty g_i^{-3}(t)\,dt < \infty, i = 1, \dots, s$*, then there exists a pressure function* p *such that* (53) *is valid.*

Remark. For the linear Stokes problem (1), (52), (53) the condition $\int_0^\infty g_i^{-3}(t)\,dt < \infty$ is not necessary.

Example 2. Domains with cylindric outlets. If $g_i(t) = 1$, then the outlet to infinity Ω_i coincides with a semicylinder $\{x \in I\!R^n : \ x' \in \sigma_i, \ x_3 > 0\}$. In this case the weights $g_i^{æ_i}(x_3)$ do not contribute to the norms of the function spaces. We have

$$\alpha_i(t) = \int_0^t g_i^{-1}(\tau)\, d\tau, \quad \kappa_{ik} = \exp(-2\beta_i t).$$

Applying the results of the previous example we get an exponential decay estimate for the solution of the Stokes problem with zero fluxes. For nonzero fluxes we have the existence results for arbitrary data as well in the linear as in the nonlinear case. These results for domains with cylindric outlets were originally obtained by O.A. Ladyzhenskaya and V.A. Solonnikov ([24]). Moreover, in [24] it is proved that for small data the solution of Navier-Stokes problem approaches as $|x| \to \infty$, $x \in \Omega_i$, the well-known Poiseuille solution (exact solution of the Navier-Stokes problem in a cylinder) which has the following form

$$u^0(x) = \frac{F_i}{\gamma_{i0}}(0, \ 0, \ u_n^0(x')), \qquad p^0(x) = \frac{F_i \nu}{\gamma_{i0}} x_n + c,$$

where u_n^0 is the solution of the problem

$$-\Delta u_n^0 = 1 \quad \text{in} \quad \sigma_i, \qquad u_n^0 = 0 \quad \text{on} \quad \partial\sigma_i$$

and

$$\gamma_{i0} = \int_{\sigma_i} u_n^0(x')\,dx' = \int_{\sigma_i} |\nabla_x u_n^0(x')|^2\,dx' > 0,$$

where $x' = (x_1, \ldots, x_{n-1})$.

Example 3. Domains with conical outlets and the aperture domain. Let us consider either the aperture domain (3) or the domain having conical outlets to infinity, i.e. $g_i(t) = g_0(1+t)$. Note that the domain (3) can be regarded as a domain with two conical outlets Ω_1 and Ω_2 of opening π. One can take $\Omega_{ik} = \{x \in \Omega_i : |x| < R_0 2^k\}$. It is easy to verify the conditions (**I**)-(**V**) and to show that the dimension of the factor-space $\widehat{H}^q(\Omega)/H^q(\Omega), q > 1$, is equal to $M - 1$. In particular, for the aperture domain (3) we have

$$\dim \widehat{H}^q(\Omega)/H^q(\Omega) = 1.$$

The divergence free vector A satisfying the flux condition (4) can be constructed just as in Example 1 (for the domain (3) the construction is even simpler, since the contour γ_{ij} can be taken to be a line).

The conditions for the solvability in $D(\Omega; r)$ of Stokes and Navier-Stokes problems can be satisfied, if we put $r_{i1} = \ldots = r_{ik} = \ldots = 1$. To get the decay properties of the Dirichlet integral over the domain ω_{ik}, we put $\kappa_{ik} = a_1^{-k}$ and we derive the following estimate

$$\|u;\ D(\omega_{ik})\| \le c 2^{-2\gamma k},$$

where γ is a sufficiently small number.

Thus, in domains with conical outlets it is possible to prove the existence of weak solutions of Stokes and Navier-Stokes problems which have finite Dirichlet integral. We formulate the corresponding results for the aperture domain (3). For details we can recommend the papers [11], [23], [53].

Theorem 14 *Let $f \in H^*(\Omega)$, F , $p_* \in \mathbb{R}^1$. Then:*
(i) *Problem* (1), (4) *has a unique weak solution $u \in \widehat{H}(\Omega)$, satisfying the estimate*

$$\|u; D(\Omega)\| \le c\,(\|f; H^*(\Omega)\| + |F|).$$

(ii) *Problem* (1), (52) *has a unique weak solution $u \in \widehat{H}(\Omega)$ satisfying the integral identity*

$$\nu \int_\Omega \nabla u \cdot \nabla \xi\,dx + p_* \int_S \xi_3\,dx' = \int_\Omega f \cdot \xi\,dx$$

for any $\xi \in \widehat{H}(\Omega)$. The corresponding estimate holds with F replaced by p_*.

Theorem 15 (i) *For arbitrary $f \in H^*(\Omega)$ and $F \in \mathbb{R}^1$ problem* (2), (4) *has at least one weak solution $u \in \widehat{H}(\Omega)$ satisfying the estimate*

$$\|u;\ D(\Omega)\| \le c\,(\|f;\ H^*(\Omega)\| + |F|).$$

(ii) *For arbitrary* $f \in H^*(\Omega)$ *and* $p_* \in I\!R^1$ *problem* (2), (52) *has at least one weak solution* $u \in \widehat{H}(\Omega)$ *and there holds the estimate*

$$\|u;\, D(\Omega)\| \le c\,(\|f;\, H^*(\Omega)\| + |p_*|).$$

Example 4. Layer-like outlet to infinity. Let $\Omega \subset I\!R^3$ have an outlet to infinity Ω_i defined by the relations

$$h^{-1}(|x'|)x_3 \in T, \quad |x'| = (x_1^2 + x_2^2)^{1/2} \ge R_0, \tag{55}$$

where $T = (t_-,\, t_+)$ is a segment in $I\!R^1$ containing the origin and $h(\tau)$ satisfies the conditions

$$c_1 h(t) \le \max_{\tau \in (t,\, 2t)} h(\tau) \le c_2 h(t), \quad h(t) \ge h_0 > 0, \tag{56}$$

$$|h(\tau) - h(\tau')| \le c_3 h(t) t^{-1} |\tau - \tau'| \quad \forall\, \tau, \tau' \in (t,\, 2t). \tag{57}$$

We put $\Omega_{ik} = \{x \in \Omega_i :\, |x'| < R_k = 2^k R_0\}$. The cut-off functions χ_{ik} can be taken to satisfy the estimates $|\nabla \chi_{ik}| \le c2^{-k}$. It can be proved (see [43]) that the constant in the estimate (9) for the domain $\omega_{ik} = \Omega_{ik+1} \setminus \Omega_{ik}$ is not greater than $c2^k R_0 h^{-1}(2^k R_0)$, where c is independent of k, and in general is not uniformly bounded. Nevertheless, in virtue of the estimate for $|\nabla \chi_{ik}|$ we get that the norm of the operator P_{ik} is uniformly bounded. It is shown in [43] that the necessary and sufficient conditions for the existence of a divergence free vector field A belonging to $D^q(\Omega_i)$, vanishing on $\partial\Omega_i \cap \partial\Omega$ and having a nonzero flux through Ω_i (i.e. $\int_{\sigma_i(t)} A \cdot n\, ds \ne 0$, where $\sigma_i(t) = \{x \in \Omega_i :\, |x'| = t\}$), is the boundedness of the integral

$$\int_1^\infty h^{1-2q}(t) t^{1-q}\, ds. \tag{58}$$

Example 5. Domain with one layer-like and one paraboloid-like outlet. The results for this example are based on the author's paper [43]. Let $\Omega \subset I\!R^3$ be a domain with two outlets to infinity Ω_1 and Ω_2 defined by the relations (55) and (31), respectively (we assume that the coordinates $\{x\}$ are introduced globally in Ω and that $t_- = 0$). The divergence free vector A satisfying the flux condition (4) may be defined by the formula

$$A(x, \varepsilon) = \mathrm{curl}(\zeta(x, \varepsilon)\, b(x)) = \nabla\zeta(x, \varepsilon) \times b(x),$$

where

$$b(x) = \frac{F}{2\pi}\Big(-\frac{x_2}{|x'|^2}, \frac{x_1}{|x'|^2}, 0\Big) \quad \text{and} \quad \zeta(x, \varepsilon) = \psi\Big(\varepsilon \ln \frac{\rho(\delta(x))}{\Delta(x)}\Big),$$

ψ and ρ are the same as in Example 1, $\Delta(x)$ and $\delta(x)$ are the regularized distances from x to $\partial\Omega \setminus \Sigma$, $\Sigma = \{x \in \partial\Omega :\, x_3 = 0\}$, and $\Sigma \cup \{x :\, |x'| = 0\}$, respectively. In Ω_2 the vector A obeys the estimates (37) and in Ω_1 one gets

$$|A(x)| \le \frac{c}{h_1(|x'|)|x'|}, \quad |\nabla A(x)| \le \frac{c}{h_1^2(|x'|)|x'|}.$$

There holds

Theorem 16 *Let $f = 0$. For any F the problem* (2), (4) *has at least one weak solution u satisfying the estimate*

$$\int_{\Omega_{(k)}} |\nabla u|^2\,dx \le c\Big(\int_1^{R_{1k}} h_1^{-3}(t)\,t^{-1}\,dt + \int_1^{R_{2k}} g_2^{-4}(t)\,dt\Big).$$

Let us assume now that $h_1(t) = 1$, $g_2(t) = 1$, i.e. for large $|x|$ the outlet Ω_1 coincides with the layer $\mathbf{L} = \{x \in I\!\!R^3 :\ 0 < x_3 < 1\}$ and the outlet Ω_2 coincides with the cylinder $\mathbf{\Pi} = \{x \in I\!\!R^3 :\ x' \in \sigma_2,\ \sigma_2 \subset I\!\!R^2,\ x_3 \in I\!\!R^1\}$. Let $(u^{02},\ p^{02})$ be a Poiseuille solution in $\mathbf{\Pi}$ defined in Example 2 and let $(u^{01},\ p^{01})$ be an exact solution of the linear Stokes problem (1) in the layer $\mathbf{L}$. In cylindrical coordinates (r, φ, z) the solution $(u^{01},\ p^{01})$ is given by the formulas

$$u^{01} = (u_r^{01}, u_\varphi^{01}, u_z^{01}) = \Big(\frac{3F(z^2 - z)}{\pi r},\ 0,\ 0\Big), \quad p^{01} = \frac{F\nu}{2\pi}\ln r. \tag{59}$$

It is proved in [43] that for sufficiently small $|F|$ the weak solution u of the nonlinear Navier-Stokes problem (2), (4) approaches as $|x| \to \infty, x \in \Omega_i, i = 1, 2$, the corresponding exact solution u^{0i} in the sense that

$$\int_{\Omega_i} |\nabla(u - u^{0i})|^2\,dx < \infty, \quad i = 1, 2.$$

2.6. Open problems

Problem 1. Theorem 1 states the coincidence of the spaces $\widehat{H}^q(\Omega)$ and $H^q(\Omega)$ for bounded and exterior domains with Lipschitz boundaries. It would be interesting either to get the same result for arbitrary domains Ω without any restrictions on $\partial\Omega$ or to construct a counterexample.

Problem 2. The solvability of the nonlinear Navier-Stokes problem (2), (4) and the estimate (49) for its solution were proved in [24], [55] under the conditions (46), (47) (see also the condition (30) for the general case). It seems to me that (46), (47) have technical nature. It would be good to get rid of these restrictions.

Problem 3. (Leray problem) Professor Leray proposed the following problem: to find a solution of the Navier-Stokes problem (2), (4) in a perturbed pipe (i.e. in the domain with two cylindric outlets to infinity) which approaches along the end of the pipe the corresponding Poiseuille flow. As we mentioned in Example 2 this question was positively answered in [24] for sufficiently small flux[2] F. Moreover, in [24] the solvability of (2), (4) was proved in the case of arbitrary flux F. However, it is not

[2] In this context see also the papers [3], [4].

known if for large data the solution approaches the Poiseuille flow and, hence, for large data the problem of Leray is still open. In [24] O.A. Ladyzhenskaya and V.A. Solonnikov reformulated Leray's problem and proposed to describe a set of all solutions to (2), (4) in a pipe, having a prescribed flux F, and to show that the solution in the perturbed pipe (which exists due to Theorem 12) tends to one of the elements of this set.

In [18] L.V. Kapitanskii has studied the problem (2), (4) in a pipe with periodically (along x_3-axis) changing section. In this case he proved the existence of a solution in a class of periodical with respect to x_3 functions for arbitrary data. In a straight pipe the section is a constant. Thus, we can consider it as a domain with periodically changing section and the period can be taken arbitrary. One approach to solve the formulated problem could be to compare the Poiseuille solution with the periodical solution of Kapitanskii and either to find a periodical solution different from Poiseuille flow or to show that for every flux F there exists a period T such that the corresponding periodical solution coincides with the Poiseuille one.

3. Strong solutions in domains with paraboloid-like outlets to infinity

3.1. Weighted function spaces

Let Ω be a domain with outlets to infinity having the form (31) (see Example 1). Denote by $L^q_{(æ,\beta)}(\Omega)$, $q = (q_0, q_1, \ldots, q_M)$, $\beta = (\beta_1, \ldots, \beta_M)$, the space of functions with the finite norm

$$\begin{aligned} \|f;\ L^q_{(æ,\beta)}(\Omega)\| &= \Big(\int_{\Omega_{(k_0)}} |f|^{q_0} dx\Big)^{1/q_0} \\ &+ \sum_{i=1}^{M} \Big(\int_{\Omega_i \setminus \Omega_{(k_0)}} G_i^{q_i æ_i} \exp\left(q_i \beta_i \alpha_i(x_n)\right) |f|^{q_i} dx\Big)^{1/q_i} \end{aligned}$$

and $\widetilde{L}^q_{(æ,\beta)}(\Omega)$ is the space of functions with the norm

$$\begin{aligned} \|f;\ \widetilde{L}^q_{(æ,\beta)}(\Omega)\| &= \Big(\int_{\Omega_{(k_0+1)}} |f|^{q_0} dx\Big)^{1/q_0} \\ &+ \sum_{i=1}^{M} \Big(\int_{\Omega_i \setminus \Omega_{(k_0-1)}} G_i^{q_i æ_i} \exp\left(q_i \beta_i \alpha_i(x_n)\right) |f|^{q_i} dx\Big)^{1/q_i}. \end{aligned}$$

Obviously, $\widetilde{L}^q_{(æ,\beta)}(\Omega) \subset L^q_{(æ,\beta)}(\Omega)$ and, if $q_0 = \ldots = q_M$ the spaces $\widetilde{L}^q_{(æ,\beta)}(\Omega)$ and $L^q_{(æ,\beta)}(\Omega)$ are equivalent.

Let $V^{l,q}_{(æ,\beta)}(\Omega)$, $l \geq 1$, be the completion of $C_0^\infty(\Omega)$ in the norm

$$\|f; V^{l,q}_{(æ,\beta)}(\Omega)\| = \sum_{|\alpha|=0}^{l} \|D^\alpha f;\ L^q_{(æ+|\alpha|-l,\beta)}(\Omega)\|$$

and $V^{-1,q}_{(æ,\beta)}(\Omega)$ be the space of functions f which can be represented in the form

$$f = f^{(0)} + (\operatorname{div} f^{(1)}, \ldots, \operatorname{div} f^{(n)}) \tag{60}$$

with $f^{(0)} \in L^q_{(æ+1,\beta)}(\Omega)$, $f^{(j)} \in L^q_{(æ,\beta)}(\Omega)$, $j = 1, \ldots, n$, and

$$\|f;\ V^{-1,q}_{(æ,\beta)}(\Omega)\| = \|f^{(0)};\ L^q_{(æ+1,\beta)}(\Omega)\| + \sum_{j=1}^{n} \|f^{(j)};\ L^q_{(æ,\beta)}(\Omega)\|.$$

Here $æ + b = (æ_1 + b, \ldots, æ_M + b)$.

The spaces $\widetilde{V}^{l,q}_{(æ,\beta)}(\Omega)$, $l \geq -1$, are defined by the same formulas with the only difference that $L^q_{(\cdot\ ,\beta)}(\Omega)$ in the definitions of the norms are replaced by $\widetilde{L}^q_{(\cdot\ ,\beta)}(\Omega)$.

The weighted Hölder space $C^{l,\delta}_{(æ,\beta)}(\Omega), l \geq 0, 1 > \delta > 0$, consists of functions u, continuously differentiable up to the order l in Ω, for which the norm

$$\begin{aligned}\|u; C^{l,\delta}_{(æ,\beta)}(\Omega)\| &= \|u; C^{l,\delta}(\Omega_{(k_0)})\| \\ &+ \sum_{i=1}^{M} \sum_{|\alpha| \leq l} \sup_{x \in \Omega_i} \{g_i^{æ_i - l - \delta + |\alpha|}(x_n) \exp(\beta_i \alpha_i(x_n)) |D^\alpha u(x)|\} \\ &+ \sum_{i=1}^{M} \sum_{|\alpha| = l} \sup_{x \in \Omega_i} \{g_i^{æ_i}(x_n) \exp(\beta_i \alpha_i(x_n)) [D^\alpha u]_\delta(x)\},\end{aligned}$$

$$\text{where} \quad [v]_\delta(x) = \sup_{y \in \Omega_i} \frac{|v(x) - v(y)|}{|x - y|^\delta},$$

is finite. Weighted function spaces in subdomains Ω' of Ω can analogously be defined. There holds the following result:

Theorem 17 ([50]) *Let $u \in V^{l,q_i}_{(æ_i,\beta_i)}(\Omega_i), \beta \in I\!R^1$.*
(i) *If s_i satisfies the conditions*

$$q_i l \leq n, \quad q_i \leq s_i \leq n q_i/(n - q_i l),$$

then $u \in L^{s_i}_{(æ_i - l + 3/q_i - 3/s_i, \beta_i)}(\Omega_i)$ and

$$\|u;\ L^{s_i}_{(æ_i - l + 3/q_i - 3/s_i, \beta_i)}(\Omega_i)\| \leq c\, \|u;\ V^{l,q_i}_{(æ_i,\beta_i)}(\Omega_i)\|. \tag{61}$$

(ii) *If the conditions*

$$q_i l > n, \quad m + \delta \leq (q_i l - n)/q_i, \quad \delta \in (0,\ 1),$$

are fulfilled, then $u \in C^{m,\delta}_{(m+\delta-l+æ_i+3/q_i,\beta_i)}(\Omega_i)$ and

$$\|u;\ C^{m,\delta}_{(m+\delta-l+æ_i+3/q_i,\beta_i)}(\Omega_i)\| \leq c\, \|u;\ V^{l,q_i}_{(æ_i,\beta_i)}(\Omega_i)\|. \tag{62}$$

The proof of Theorem 17 is based on classical embedding results and on simple scaling arguments applied in the domains ω_{ik}.

3.2. Stokes problem. Solutions with zero fluxes

Let us consider the Stokes problem (1), (4) in Ω. Below we denote the problem (1), (4) with zero fluxes, i.e. $F_i = 0$, $i = 1, \dots, M$, by (1), $(4)_0$.

Using the well-known local estimates for elliptic problems [2], [52] and the scaling argument in domains ω_{il}, one can prove the following weighted local estimates. Thereby $u \in B_{\text{loc}}(\Omega)$ means that $u \in B(\Omega')$ for arbitrary bounded subdomains $\Omega' \subset \Omega$.

Theorem 18 (i) *Let the function f have the representation (60) and $f^{(0)}$, $f^{(j)} \in L^{q_i}_{\text{loc}}(\Omega_i)$, $j = 1, 2, 3$, $q_i \geq 2$, $i = 0, 1, \dots, M$. Then the weak solution u of the Stokes problem (1), $(4)_0$ satisfies the local estimate*

$$\|\nabla u;\ L^{q_i}_{(æ_i,\beta_i)}(\omega_{is})\| \leq c\Big(\|f^{(0)};\ L^{q_i}_{(æ_i+1,\beta_i)}(\omega^*_{is})\| + \sum_{j=1}^{n}\|f^{(j)};\ L^{q_i}_{(æ_i,\beta_i)}(\omega^*_{is})\| + \|\nabla u;\ L^{2}_{(æ_i+\widehat{æ}_i,\beta_i)}(\omega^*_{is})\|\Big), \tag{63}$$

where $\widehat{æ}_i = n(2-q_i)/2q_i$, $\omega^*_{is} = \omega_{is-1} \bigcup \omega_{is} \bigcup \omega_{is+1}$.

(ii) *Assume that $\partial\Omega \in C^{l+2}$, $f \in W^{l,q_i}_{\text{loc}}(\Omega_i)$, $q_i > 1$, $l \geq 0$. Then the solution (u,p) of problem (1), $(4)_0$ satisfies the local estimates*

$$\|u;\ V^{l+2,q_i}_{(æ_i,\beta_i)}(\omega_{is})\| + \|\nabla p;\ V^{l,q_i}_{(æ_i,\beta_i)}(\omega_{is})\| \leq c\Big(\|f;\ V^{l,q_i}_{(æ_i,\beta_i)}(\omega^*_{is})\| + \|\nabla u;\ L^{q_i}_{(æ_i-l-1,\beta_i)}(\omega^*_{is})\|\Big), \tag{64}$$

$$\|u;\ V^{l+2,q_i}_{(æ_i,\beta_i)}(\omega_{is})\| + \|\nabla p;\ V^{l,q_i}_{(æ_i,\beta_i)}(\omega_{is})\| \leq c\Big(\|f;\ V^{l,q_i}_{(æ_i,\beta_i)}(\omega^*_{is})\| + \|\nabla u;\ L^{2_i}_{(æ_i-l-n/2-1+n/q_i,\beta_i)}(\omega^*_{is})\|\Big). \tag{65}$$

(iii) *Assume that $\partial\Omega \in C^{l+2,\delta}$ and let $f \in C^{l,\delta}_{\text{loc}}(\Omega_i)$, $l \geq 0$, $0 < \delta < 1$. Then the solution (u,p) of (1), $(4)_0$ satisfies the local estimate*

$$\|u;\ C^{l+2,\delta}_{(æ_i,\beta_i)}(\omega_{is})\| + \|\nabla p;\ C^{l,\delta}_{(æ_i,\beta_i)}(\omega_{is})\| \leq c\Big(\|f;\ C^{l,\delta}_{(æ_i,\beta_i)}(\omega^*_{is})\| + \|\nabla u;\ L^{2}_{(æ_i-l-1-n/2-\delta,\beta_i)}(\omega^*_{is})\|\Big). \tag{66}$$

The constants are independent of s and f.

Let us formulate the main results of the section.

Theorem 19 (i) [48] *Assume*[3] *that $\partial\Omega \in C^{l+2}$, $f \in \widetilde{V}^{l,q}_{(æ,\beta)}(\Omega)$, $l \geq -1$, $q_j > 1$, $|\beta_j| < \beta_*$, æ is arbitrary. Then there exists a unique*[4] *solution (u,p) of problem (1), $(4)_0$ with $u \in V^{l+2,q}_{(æ,\beta)}(\Omega)$ and*

$$\|u;\ V^{l+2,q}_{(æ,\beta)}(\Omega)\| \leq c\|f;\ \widetilde{V}^{l,q}_{(æ,\beta)}(\Omega)\|. \tag{67}$$

---

[3]If $l = -1$, it is enough to assume that $\partial\Omega$ is only Lipschitz.

[4]Speaking about the uniqueness of the solution (u,p), we have in mind that the pressure p is unique up to an additive constant.

Moreover, if $l \geq 0$, *we have* $\nabla p \in V^{l,q}_{(æ,\beta)}(\Omega)$ *and there holds the estimate*

$$\|u;\ V^{l+2,q}_{(æ,\beta)}(\Omega)\| + \|\nabla p;\ V^{l,q}_{(æ,\beta)}(\Omega)\| \leq c\|f;\ \widetilde{V}^{l,q}_{(æ,\beta)}(\Omega)\|. \tag{68}$$

(ii) [49] *Assume that* $\partial\Omega \in C^{l+2,\delta}$ *and* $f \in C^{l,\delta}_{(æ,\beta)}(\Omega)$, *where* $l \geq 0$, $\delta \in (0,1)$, $|\beta_i| < \beta_*$ *and* æ *is arbitrary. Then problem* (1), $(4)_0$ *has a unique solution* (u,p) *with* $u \in C^{l+2,\delta}_{(æ,\beta)}(\Omega)$, $\nabla p \in C^{l,\delta}_{(æ,\beta)}(\Omega)$. *Moreover, there holds the estimate*

$$\|u;\ C^{l+2,\delta}_{(æ,\beta)}(\Omega)\| + \|\nabla p;\ C^{l,\delta}_{(æ,\beta)}(\Omega)\| \leq c\|f;\ C^{l,\delta}_{(æ,\beta)}(\Omega)\|. \tag{69}$$

Remark. In particular, if $f \in C^{l,\delta}_{(æ,\beta)}(\Omega)$ with $\beta_i > 0$ (for example, f has a compact support), then from Theorem 19(ii) there follow the exponential decay estimates for the solution $(u,\ p)$ of problem (1), $(4)_0$, i.e. there holds for $x \in \Omega_i$,

$$\begin{aligned}
|D^\alpha u(x)| &\leq c\exp\Big(-\beta_i\int_0^{x_n^{(i)}} g_i^{-1}(t)dt\Big) g_i^{l+2+\delta-|\alpha|-æ_i}(x_n^{(i)}), \quad 0 \leq |\alpha| \leq l+2,\\
|D^\alpha \nabla p(x)| &\leq c\exp\Big(-\beta_i\int_0^{x_n^{(i)}} g_i^{-1}(t)dt\Big) g_i^{l+\delta-|\alpha|-æ_i}(x_n^{(i)}), \quad 0 \leq |\alpha| \leq l.
\end{aligned}$$

We discuss shortly the most important steps of the proof. The detailed calculations can be found in [48], [49].

Step 1. First of all, we consider the case, when $q_i \geq 2$ for all $i \in \{0,1,\dots,M\}$. In the domain Ω we introduce the partition of unity $\{\varphi_{k_0}, \varphi_{ik}\}$, $i = 1,\dots,M$, $k = k_0+1,\dots$, subordinated to the covering of Ω by the domains $\{\Omega_{(k_0+1)}, \omega^*_{ik},\}$, $i = 1,\dots,M$, $k = k_0+1,\dots$, i.e. $\operatorname{supp}\varphi_{k_0} \subset \bar{\Omega}_{(k_0+1)}$, $\operatorname{supp}\varphi_{ik} \subset \bar{\omega}^*_{ik}$, $\varphi_{k_0}, \varphi_{ik} \in C_0^\infty(I\!R^n)$,

$$\Omega = \Omega_{(k_0+1)} \bigcup \Big(\bigcup_{i=1}^{M}\Big(\bigcup_{k=k_0+1}^{\infty} \omega^*_{ik}\Big)\Big)$$

and

$$\sum_{i=1}^{M}\sum_{k=k_0+1}^{\infty} \varphi_{ik}(x) + \varphi_{k_0}(x) = 1 \quad \text{in } \Omega.$$

Let

$$\begin{aligned}
f^{(j)}_{k_0} &= \varphi_{k_0} f^{(j)}, \qquad f^{(j)}_{ik} = \varphi_{ik} f^{(j)}, \quad j = 0,1,\dots,n-1,\\
f_{k_0} &= f^{(0)}_{k_0} + (\operatorname{div} f^{(1)}_{k_0},\ \dots,\ \operatorname{div} f^{(n)}_{k_0}),\\
f_{ik} &= f^{(0)}_{ik} + (\operatorname{div} f^{(1)}_{ik},\ \dots,\ \operatorname{div} f^{(n)}_{ik}),\\
f^{[N]}_i &= \sum_{k=k_0+1}^{N} f_{ik},\ \ f^{[N]} = f_{k_0} + \sum_{i=1}^{M} f^{[N]}_i.
\end{aligned}$$

Obviously, the functions f_{k_0}, f_{ik} have compact supports:

$$\operatorname{supp} f_{k_0} \subset \bar{\Omega}_{k_0}, \ \operatorname{supp} f_{ik} \subset \bar{\omega}^*_{ik}.$$

Then for each f_{k_0}, f_{ik} there exists a unique weak solution u_{k_0}, u_{ik} of problem (1), $(4)_0$ with finite Dirichlet integral (see Section 2).

Step 2. Applying the difference inequality techniques (Saint–Venant's principle) we prove that each u_{ik} admits the estimate

$$\begin{aligned} \|\nabla u_{ik};\ L^{q_i}_{(æ_i,\beta_i)}(\omega^*_{il})\| &\le c\exp\big(-\varepsilon_0 c_0|k-l|\big)\|f_{ik};\ V^{-1,2}_{(æ_i+\widehat{æ}_i,\beta_i)}(\omega^*_{il})\| \\ &\le c\exp\big(-\varepsilon_0 c_0|k-l|\big)\|f_{ik};\ V^{-1,q_i}_{(æ_i,\beta_i)}(\omega^*_{il})\|, \end{aligned} \tag{70}$$

where $\widehat{æ}_i = n(2-q_i)/2q_i$ and $|\beta_i| < \beta_*$.

Step 3. Let

$$u^{[N]} = \sum_{i=1}^{M} \sum_{k=k_0+1}^{N} u_{ik} + u_{k_0}.$$

The repeated application of (70), Hölder and triangle inequalities allow us to prove that $u^{[N]} \in V^{1,q}_{(æ,\beta)}(\Omega)$ and

$$\|u^{[N]};\ V^{1,q}_{(æ,\beta)}(\Omega)\| \le c\|f^{[N]};\ \widetilde{V}^{-1,q}_{(æ,\beta)}(\Omega)\| \le c\|f;\ \widetilde{V}^{-1,q}_{(æ,\beta)}(\Omega)\|.$$

Therefore, the sequence $\{u^{[N]}\}$ converges in the norm of the space $V^{1,q}_{(æ,\beta)}(\Omega)$ to the solution $u \in V^{1,q}_{(æ,\beta)}(\Omega)$ of problem (1), $(4)_0$.

Step 4. Applying to $\{u^{[N]}\}$ the weighted local estimates (63)-(65), we conclude that $u \in V^{l+2,q}_{(æ,\beta)}(\Omega)$, $\nabla p \in V^{l+2,q}_{(æ,\beta)}(\Omega)$ and obtain the estimates (67), (68) for u and ∇p.

Step 5. The proof of the case where $q_i < 2$ for certain $i \in \{0,1,\dots,M\}$ follows now by duality arguments.

Step 6. The weighted Hölder solvability of problem (1), $(4)_0$ is proved by the same scheme with the only difference that the local estimate (66) is applied in place of (63)-(65).

3.3. Stokes problem. Solutions with nonzero fluxes

Let us consider the Stokes problem (1), (4) with arbitrary nonzero fluxes F_i, $i = 1,\dots,M$. We assume that for each $i \in \{1,\dots,M\}$ there exists a number q^*_i such that

$$\int_0^\infty g_i^{-(n-1)(q^*_i-1)-q^*_i}(t)\,dt < \infty. \tag{71}$$

We look for the solution u in the form

$$u = A + v, \tag{72}$$

where A is the divergence free vector field, satisfying the flux condition (4) and estimate (37) (or (38), if $n = 2$), and v is the solution of the Stokes problem with zero fluxes (1), $(4)_0$, corresponding to the right–hand side $f + \nu\Delta A$. In virtue of (37), (38) it is easy to verify that there hold the estimates

$$\|A;\ \widehat{H}^{q^*}(\Omega)\| \ \leq\ c\sum_{i=1}^{M}|F_i|\Big(\int_0^\infty g_i^{-(n-1)(q_i^*-1)-q_i^*}(t)dt\Big)^{1/q_i^*} \leq c\sum_{i=1}^{M}|F_i|, \quad (73)$$

$$\|\Delta A;\ \widetilde{V}^{l,q}_{(æ^*,0)}(\Omega)\| \ \leq\ c\sum_{i=1}^{M}|F_i|, \quad (74)$$

$$\|A;\ C^{l+2,\delta}_{(æ,0)}(\Omega)\| \ \leq\ c\sum_{i=1}^{M}|F_i|, \quad (75)$$

where for $i = 1, \dots, M$

$$æ^* = (æ_1^*, \dots, æ_M^*), \qquad æ_i^* = n + 1 + l - nq_i^*/q_i,$$
$$æ = (æ_1, \dots, æ_M), \qquad æ_i = n + 1 + l + \delta. \quad (76)$$

Using these estimates and applying Theorem 19 to the solution $(v,\ p)$ of problem (1), $(4)_0$, we derive the following results.

Theorem 20 (i) [48] *Let the condition (71) hold. Then for arbitrary* $f \in \widetilde{V}^{-1,q^*}_{(0,0)}(\Omega)$ *and* $F_i \in I\!R^1$, $i = 1, \dots, M$, *problem (1), (4) has a unique solution* $u \in \widehat{H}^{q^*}(\Omega)$ *satisfying the estimate*

$$\|u;\ \widehat{H}^{q^*}(\Omega)\| \leq c\Big(\|f;\ \widetilde{V}^{-1,q^*}_{(0,0)}(\Omega)\| + \sum_{i=1}^{M}|F_i|\Big). \quad (77)$$

(ii) [48] *Let* $\partial\Omega \in C^{l+2}$, $F_i \in I\!R^1$, $i = 1, \dots, M$, $f \in \widetilde{V}^{l,q}_{(æ^*,0)}(\Omega)$, *where* $l \geq -1$, $q_i > 1$ *and* $æ_i^*$ *are defined by the formula (76). Then there exists a unique solution* (u, p) *of problem (1), (4) with* $u \in V^{l+2,q}_{(æ^*,0)}(\Omega)$, $\nabla p \in V^{l,q}_{(æ^*,0)}(\Omega)$ *and*

$$\|u;\ V^{l+2,q}_{(æ^*,0)}(\Omega)\| + \|\nabla p;\ V^{l,q}_{(æ^*,0)}(\Omega)\| \leq c\Big(\sum_{i=1}^{M}|F_i| + \|f;\ \widetilde{V}^{l,q}_{(æ^*,0)}(\Omega)\|\Big). \quad (78)$$

In particular, if $f \in V^{-1,q}_{(æ^*,0)}(\Omega)$, *then* $u \in \widehat{H}^{1,q}_{æ^*}(\Omega)$.
(iii) [49] *Let* $\partial\Omega \in C^{l+2,\delta}$, $f \in C^{l,\delta}_{(æ,0)}(\Omega)$, $l \geq 0$, $\delta \in (0,1)$, *where* $æ$ *is defined by (76). Then there exists a unique solution* (u, p) *of problem (1), (4) such that* $u \in C^{l+2,\delta}_{(æ,0)}(\Omega)$, $\nabla p \in C^{l,\delta}_{(æ,0)}(\Omega)$ *and there holds the estimate*

$$\|u;\ C^{l+2,\delta}_{(æ,0)}(\Omega)\| + \|\nabla p;\ C^{l,\delta}_{(æ,0)}(\Omega)\| \leq c\Big(\sum_{i=1}^{M}|F_i| + \|f;\ C^{l,\delta}_{(æ,0)}(\Omega)\|\Big). \quad (79)$$

In particular, from (79) it follows that

$$|D^\alpha u(x)| \;\leq\; c\Big(\sum_{i=1}^{M}|F_i| + \|f;\; C^{l,\delta}_{(æ,0)}(\Omega)\|\Big) g_i^{-n+1-|\alpha|}(x_n), \tag{80}$$

$$|D^\alpha \nabla p(x)| \;\leq\; c\Big(\sum_{i=1}^{M}|F_i| + \|f;\; C^{l,\delta}_{(æ,0)}(\Omega)\|\Big) g_i^{-n-1-|\alpha|}(x_n) \tag{81}$$

for $x \in \Omega_i$, $|\alpha| \geq 0$. *Moreover, for* $x \in \Omega_i$

$$|p(x)| \leq c\Big(\sum_{i=1}^{M}|F_i| + \|f;\; C^{l,\delta}_{(æ,0)}(\Omega)\|\Big)\Big(\int_0^{x_n} g_i^{-n-1}(t)dt\Big) + c_1. \tag{82}$$

Concerning the proof of this theorem, we should mention additionally, that in order to prove (82), we represent the function $p(x)$ in the form

$$p(x) = p(x_0) + \int_{x_0}^{x} \frac{\partial p}{\partial \gamma}\, d\gamma,$$

where $\gamma \subset \Omega_i$ is a smooth contour, connecting some fixed point $x_0 \in \Omega_i$ with an arbitrary point $x \in \Omega_i$, such that γ is given by the equations $x_j = \gamma_j(x_n)$, $j = 1, \ldots, n-1$ and $(1 + \gamma_1'(x_n)^2 + \ldots + \gamma_{n-1}'(x_n)^2)^{1/2} \leq \text{const}$. Then,

$$\begin{aligned} |p(x)| \;&\leq\; |p(x_0)| + \Big|\int_{x_0}^{x} \frac{\partial p}{\partial \gamma}\, d\gamma\Big| \leq |p(x_0)| \\ &\quad + \sup_{x\in\Omega_i} |\nabla p(x) g_i^{n+1}(x_n)| \int_{x_{0n}}^{x_n} g_i^{-n-1}(t)\sqrt{1 + \gamma_1'(t)^2 + \ldots + \gamma_{n-1}'(t)^2}dt \\ &\leq\; |p(x_0)| + c\|\nabla p;\; C^{l,\delta}_{(æ,0)}(\Omega)\| \int_0^{x_n} g_i^{-n-1}(t)\, dt. \end{aligned}$$

Remark. The estimates for the pressure $p(x)$ in domains with outlets to infinity have been obtained by V.A. Solonnikov (1981, [53]). It is proved in [53] that for each outlet Ω_i with $\int_0^\infty g_i^{-n-1}(t)\, dt < \infty$ the pressure $p(x)$ tends to a constant p_i as $|x| \to \infty$, $x \in \Omega_i$. Estimate (82) gives in this case $|p(x)| \leq \text{const.}$, which agrees with the results from [53].

3.4. Navier–Stokes problem

Let us consider the nonlinear Navier–Stokes system (2), (4). The problem (2), (4) with zero fluxes, i.e. $F_i = 0$, $i = 1, \ldots, M$, is denoted by (2), $(4)_0$.

The Navier–Stokes problem with zero fluxes is studied in weighted Sobolev and Hölder spaces $V^{l,q}_{(æ,\beta)}(\Omega)$ and $C^{l,\delta}_{(æ,\beta)}(\Omega)$, $\beta_i > 0$, $i = 1, \ldots, M$. These functions decay exponentially at infinity. The solvability is proved for small data.

Theorem 21 ([50]) (i) *Let* $\partial\Omega \in C^{l+2}$ *and* $f \in \widetilde{V}^{l,q}_{(æ,\beta)}(\Omega)$ *with*

$$l \geq 0, \;\; q_i > 1, \;\; \beta_* > \beta_i > 0, \;\; æ_i \;\; is \;\; arbitrary. \tag{83}$$

Then for sufficiently small $\|f;\ \widetilde{V}^{l,q}_{(æ,\beta)}(\Omega)\|$ *problem* (2), $(4)_0$ *has a unique solution* $(u,\ p)$ *with* $u \in V^{l+2,q}_{(æ,\beta)}(\Omega)$, $\nabla p \in V^{l,q}_{(æ,\beta)}(\Omega)$ *and the following estimate holds true*

$$\|u;\ V^{l+2,q}_{(æ,\beta)}(\Omega)\| + \|\nabla p;\ V^{l,q}_{(æ,\beta)}(\Omega)\| \leq c\|f;\ \widetilde{V}^{l,q}_{(æ,\beta)}(\Omega)\|. \tag{84}$$

(ii) *Let* $\partial\Omega \in C^{l+2,\delta}$ *and* $f \in C^{l,\delta}_{(æ,\beta)}(\Omega)$,

$$l \geq 0, \;\; 1 > \delta > 0, \;\; \beta_* > \beta_i > 0, \;\; æ_i \;\; is \;\; arbitrary. \tag{85}$$

If the norm $\|f;\ C^{l,\delta}_{(æ,\beta)}(\Omega)\|$ *is sufficiently small, then* (2), $(4)_0$ *has a unique solution* $(u,\ p)$ *with* $u \in C^{l+2,\delta}_{(æ,\beta)}(\Omega)$, $\nabla p \in C^{l,\delta}_{(æ,\beta)}(\Omega)$ *and*

$$\|u;\ C^{l+2,\delta}_{(æ,\beta)}(\Omega)\| + \|\nabla p;\ C^{l,\delta}_{(æ,\beta)}(\Omega)\| \leq c\|f;\ C^{l,\delta}_{(æ,\beta)}(\Omega)\|. \tag{86}$$

To prove the statements of the theorem we put the nonlinear term $(u\cdot\nabla)u$ to the right and we consider the nonlinear problem (2), $(4)_0$ as the linear one with the right–hand side equal to $f-(u\cdot\nabla)u$. Applying Theorem 19, we reduce the problem to the operator equation either in the space $V^{l+2,q}_{(æ,\beta)}(\Omega)$ or in the space $C^{l+2,\delta}_{(æ,\beta)}(\Omega)$:

$$u = \mathcal{A}u.$$

By using the weighted embedding theorem we prove that for sufficiently small data the nonlinear terms define a contraction operator in a small ball of the spaces $V^{l+2,q}_{(æ,\beta)}(\Omega)$ and $C^{l+2,\delta}_{(æ,\beta)}(\Omega)$. Therefore, the claims of the theorem follow from the Banach contraction principle.

The main result for the problem (2), (4) with nonzero fluxes F_i, $i = 1,\dots,M$, reads as follows:

Theorem 22 ([50]) *Let* $\Omega \subset I\!R^3$ *be a domain with* $M > 1$ *outlets to infinity. Assume that, in addition to* (32), (33) *the functions* g_i *satisfy the conditions* (46), (47). *Let* $\partial\Omega \in C^{l+2,\delta}$, $l \geq 0$, $0 < \delta < 1$, $f = 0$ *and suppose that for each* $i \in \{1,\dots,M\}$ *there exists a number* $q_i^* \geq 3/2$ *such that*

$$\int_0^\infty g_i^{-3q_i^*+2}(t)\,dt < \infty. \tag{87}$$

(i) *Then for arbitrary fluxes* $F_i \in I\!R^1, i = 1,\dots,M$, *there exists a solution* u *of problem* (2), (4), *belonging to the space* $V^{l+2,q}_{(æ^*,0)}(\Omega)$ *with*

$$æ_i^* = 4 + l - 3q_i^*/q_i, \quad i = 1,\dots,M. \tag{88}$$

Moreover, there exists the pressure function $\nabla p \in V^{l,q}_{(æ^*,0)}(\Omega)$ *and the following estimate holds true*

$$\|u;\ V^{l+2,q}_{(æ^*,0)}(\Omega)\| + \|\nabla p;\ V^{l,q}_{(æ^*,0)}(\Omega)\| \le C(|F|). \tag{89}$$

In particular, $u \in \widehat{H}^q_{æ^*}(\Omega)$ $(u \in \widehat{H}^{q^*}(\Omega))$.

(ii) *The solution* (u, p) *of problem* (2), (4) *admits the pointwise estimates*

$$|D^\alpha u(x)| \le C(|F|) g_i^{-2-|\alpha|}(x_3), \qquad x \in \Omega_i, \quad 0 \le |\alpha| \le l, \tag{90}$$

$$|D^\alpha \nabla p(x)| \le C(|F|) g_i^{-3-|\alpha|}(x_3), \qquad x \in \Omega_i, \quad 0 < |\alpha| \le l, \tag{91}$$

$$|p(x)| \le C(|F|) \int_0^{x_3} g_i^{-4}(t)\, dt + c_1, \quad x \in \Omega_i. \tag{92}$$

The proof of this theorem is based on the estimate (49) obtained for a weak solution of the problem with unbounded Dirichlet integral.

On the heuristical level one can see that each function v satisfying estimates (49) and admitting the decay estimates

$$|D^\alpha v(x)| \le c\, g_i^{\gamma-|\alpha|}(x_3), \quad x \in \Omega_i,\ |\alpha| \ge 0,$$

should have, minimally, the decay rate $v \sim g_i^{-11/6-|\alpha|}(x_3)$, i.e. $\gamma = -11/6$. Then

$$\Delta v \sim g_i^{-23/6}(x_3), \quad (v\cdot\nabla)v \sim g_i^{-28/6}(x_3).$$

Thus, the nonlinear term $(v \cdot \nabla)v$ decays at infinity faster than the linear one Δv and by bootstrap arguments we can improve the estimates for v and derive (89)-(92). Having this simple idea as background, Theorem 22 is proved by repeated application of weighted embedding theorems, local weighted estimates and results on the linear Stokes problem (1), $(4)_0$ (see Theorem 19).

For the two-dimensional domain Ω the analogous result cannot be proved by the same method. This is related to the fact that in this case the divergence free vector A satisfying the flux conditon (4), has a decay rate as $g_i^{-1}(x_2)$ and hence,

$$\Delta A \sim g_i^{-3}(x_2), \quad (A\cdot\nabla)A \sim g_i^{-3}(x_2) \quad \text{as } |x| \to \infty,\ x \in \Omega_i.$$

Thus, the linear and nonlinear terms have the same power at infinity. For domains $\Omega \subset I\!R^2$ with $M > 1$ we have the results only for small data. The proof of the following theorem is based on the Banach contraction principle.

Theorem 23 ([50]) *Let* $\Omega \subset I\!R^2$ *be a domain with* $M > 1$ *outlets to infinity. For sufficiently small* $|F|$ *problem* (2), (4) *has a unique solution* $(u,\ p)$ *satisfying representation* (72) *and for* $x \in \Omega_i$, $0 \le |\alpha| \le l$ *the estimates*

$$|D^\alpha u(x)| \le C(|F|) g_i^{-1-|\alpha|}(x_2),$$

$$|D^\alpha \nabla p(x)| \le C(|F|) g_i^{-2-|\alpha|}(x_2),$$

$$|p(x)| \le C(|F|) \int_0^{x_2} g_i^{-3}(t)\, dt + c_1.$$

Remark. Theorems 22–23 are also valid for nonzero right–hand sides f having an appropriate decay at infinity.

3.5. Asymptotics of the solutions

We study the asymptotics of solutions to the Stokes and Navier–Stokes problems under the assumption that

$$g_i(t) = g_0 t^{1-\gamma}, \qquad 0 < \gamma < 1.$$

Above, in order to find the solutions with nonzero fluxes F_i, we were looking for the velocity field u in the form

$$u = A + v, \tag{93}$$

where A is a divergence free vector function, satisfying the flux conditions (4) and the estimate

$$|D^\alpha A(x)| \le C(|F|)\, g_i(x_n)^{-n+1-|\alpha|}. \tag{94}$$

Then for $(v,\, p)$ we derive the Stokes problem (1), $(4)_0$ (i.e. $F_i = 0, i = 1, \ldots, M$) and the new right–hand side $f + \nu\Delta A$. Finally, for the solution v of this problem, which can be considered as a perturbation of A, we get the same decay estimates as for A itself. This is related to the fact that the "flux function" A was taken arbitrary.

The formal asymptotics of the solutions were constructed by S.A. Nazarov and the author in [36], where we also proved better decay estimates for the remainder. For example, in the three–dimensional case the obtained asymptotical solution for the Stokes problem with zero right–hand side has the form

$$\begin{aligned} P^{[N]}(x) &= x_3^{\lambda_0} \sum_{k=0}^{N} x_3^{-2k\gamma}(q_k^{(0)} + x_3^{-2\gamma} Q_k(x_3^{\gamma-1}x')), \\ U_3^{[N]}(x) &= x_3^{\lambda_0-\gamma+1} \sum_{k=0}^{N} x_3^{-2(k+1)\gamma} U_{3k}(x_3^{\gamma-1}x'), \\ U_j^{[N]}(x) &= x_3^{\lambda_0-3\gamma+1} \sum_{k=0}^{N} x_3^{-(2k+1)\gamma} U_{jk}(x_3^{\gamma-1}x'),\ \ j = 1, 2, \end{aligned} \tag{95}$$

where $q_k^{(0)}$ are constants and $\lambda_0 = 4\gamma - 3$, $\gamma \ne 3/4$. If $\gamma = 3/4$, the representation for the pressure $P^{[N]}$ contains the logarithmic term. Estimates for the discrepancies which are left by this approximate solution $(U^{[N]},\ P^{[N]})$ in the Stokes equations improve when we increase N and, therefore, we get "good" decay estimates for the remainder $(v = u - U^{[N]},\ q = p - P^{[N]})$. The procedure which we use to construct the formal asymptotics is a variant of the well-known algorithm of constructing the asymptotics for solutions to elliptic equations in slender domains (e.g. S. A. Nazarov [30]). In order to explain the anology between the paraboloids and the slender domains, let us consider the intersection of Ω_i with the sphere S_R^2 of radius R. After the change of variables $x \to R^{-1}x = \xi$ the sphere S_R^2 goes over to the unit sphere S_1^2 and the intersection $\Omega_i \cap S_R^2$ turns out to become a domain with small, of order $O(R^{-\gamma})$, diameter. This property turns us to introduce the "transversal stretched coordinates"

$$\eta_j = x_3^{\gamma-1} x_j, \;\; j = 1, 2, \quad \eta_3 = x_3$$

while the image of the domain $\Omega_i \cap S_R^2$ is independent of R. After this, applying formally the methods of the theory of elliptic equations in slender domains, we derive for the pressure p the one–dimensional Reynolds equation (see S. A. Nazarov, K. Pileckas [35]), which follows as a compatibility condition for the solvability of the two–dimensional Stokes problem (in the domain $\omega = \{\eta' \in I\!R^2 \; : |\eta'| < g_0\}$) for the velocity field u.

Let us discuss the construction of an "approximate solution at infinity" of the problem (1), (4). We consider the homogeneous problem (1), (4) (i.e. $f = 0$) in the outlet to infinity

$$\Omega_i = \{x \in I\!R^n \colon \; |x'| < g_0 x_n^{1-\gamma}, \; x_n > 1\}, \quad 0 < \gamma < 1.$$

Passing to new coordinates

$$\eta_j = x_n^{\gamma-1} x_j, \;\; j = 1, \ldots, n-1, \quad \eta_n = x_n \tag{96}$$

in (1) and using the evident relations

$$\begin{aligned}
\frac{\partial}{\partial x_j} &= \eta_n^{\gamma-1} \frac{\partial}{\partial \eta_j}, \qquad \frac{\partial^2}{\partial x_j^2} = \eta_n^{2\gamma-2} \frac{\partial^2}{\partial \eta_j^2}, \quad j = 1, \ldots, n-1, \\
\frac{\partial}{\partial x_n} &= \frac{\partial}{\partial \eta_n} - \sum_{j=1}^{n-1} (1-\gamma) \eta_j \eta_n^{-1} \frac{\partial}{\partial \eta_j}, \\
\frac{\partial^2}{\partial x_n^2} &= \frac{\partial^2}{\partial \eta_n^2} + 2(\gamma-1)\eta_n^{-1} \sum_{j=1}^{n-1} \eta_j \frac{\partial^2}{\partial \eta_n \partial \eta_j} + (\gamma-1)(\gamma-2)\eta_n^{-2} \sum_{j=1}^{n-1} \eta_j \frac{\partial}{\partial \eta_j} \\
&\quad + \sum_{j,l=1}^{n-1} (\gamma-1)^2 \eta_n^{-2} \eta_l \eta_j \frac{\partial^2}{\partial \eta_l \partial \eta_j},
\end{aligned} \tag{97}$$

we rewrite (1) in the following form:

$$\begin{aligned}
-\nu\Big(\eta_n^{2\gamma-2}\Delta' + \mathcal{D}^2\Big)u' + \eta_n^{\gamma-1}\nabla' p &= 0 \quad \text{in } \Pi_+ \\
-\nu\Big(\eta_n^{2\gamma-2}\Delta' + \mathcal{D}^2\Big)u_n + \mathcal{D}p &= 0 \quad \text{in } \Pi_+ \\
\eta_n^{\gamma-1}\mathrm{div}' u' + \mathcal{D}u_n &= 0 \quad \text{in } \Pi_+, \\
u &= 0 \quad \text{on } S_+.
\end{aligned} \tag{98}$$

In (98) we have used the notations

$$\mathcal{D} = \partial_n + (\gamma-1)\eta_n^{-1}\eta' \cdot \nabla',$$

$$\begin{aligned}\Pi_+ &= \{\eta \in I\!R^n\colon\ |\eta'| < g_0,\ \eta_n > 1\}, \qquad S_+ = \{\eta \in I\!R^n\colon\ |\eta'| < g_0,\ \eta_n > 1\},\\ u' &= (u_1, \ldots, u_{n-1}), \qquad \partial_k = \partial/\partial\eta_k, \qquad k = 1, \ldots, n,\\ \nabla' &= (\partial_1, \ldots, \partial_{n-1}), \qquad \mathrm{div}' u' = \nabla' \cdot u', \qquad \Delta' = \nabla' \cdot \nabla'.\end{aligned}$$

We look for the solution (U_0, P_0) of (98) in the form

$$P_0(\eta', \eta_n) = q_0(\eta_n) + Q_0(\eta', \eta_n),$$
$$U_0(\eta', \eta_n) = (U_0'(\eta', \eta_n), U_{n0}(\eta', \eta_n)) \tag{99}$$

with

$$U_{n0}(\eta', \eta_n) = \eta_n^{2(1-\gamma)} \partial_n q_0(\eta_n) \Phi(\eta'). \tag{100}$$

Let us assume that the infinitesimals $\eta_n^k \partial_n^{k+1} q_0(\eta_n)$, $k = 1, 2, \ldots$, are equivalent (as $\eta_n \to \infty$) to $\partial_n q_0(\eta_n)$ (this assumption can be justified). Substituting (U_0, P_0) into equations (98) and selecting the leading at infinity terms, we derive

$$\begin{cases} -\nu \partial_n q_0(\eta_n) \Delta' \Phi(\eta') + \partial_n q_0 = 0 & \text{in } \omega,\\ \Phi(\eta') = 0 & \text{on } \partial\omega \end{cases}$$

and

$$\begin{cases} -\nu \eta_n^{2\gamma-2} \Delta' U_0'(\eta') + \eta_n^{\gamma-1} \nabla' Q_0(\eta') = 0 & \text{in } \omega,\\ \eta_n^{\gamma-1} \mathrm{div}' U_0'(\eta') = -\mathcal{D} U_{n0}(\eta', \eta_n) & \text{in } \omega,\\ U_0'(\eta') = 0 & \text{on } \partial\omega \end{cases}$$

or, what is the same,

$$\begin{cases} \nu \Delta' \Phi(\eta') = 1 & \text{in } \omega,\\ \Phi(\eta') = 0 & \text{on } \partial\omega, \end{cases} \tag{101}$$

$$\begin{cases} -\nu \Delta' U_0'(\eta') + \nabla'(\eta_n^{1-\gamma} Q_0(\eta')) = 0 & \text{in } \omega,\\ \mathrm{div}' U_0'(\eta') = G_0(\eta', \eta_n) & \text{in } \omega,\\ U_0'(\eta') = 0 & \text{on } \partial\omega, \end{cases} \tag{102}$$

where $\omega = \{\eta' \in I\!R^{n-1}\colon\ |\eta'| < g_0\}$,

$$G_0(\eta', \eta_n) = -\eta_n^{1-\gamma} \mathcal{D}(\eta_n^{2(1-\gamma)} \partial_n q_0(\eta_n) \Phi(\eta')). \tag{103}$$

Multiplying (101) by $\Phi(\eta')$ and integrating by parts one gets

$$\int_\omega \Phi(\eta')\, d\eta' = -\nu \int_\omega |\nabla' \Phi|^2\, d\eta' = \kappa_0 < 0.$$

The solution $\Phi(\eta')$ to (101) has the form

$$\Phi(\eta') = \frac{1}{2\nu(n-1)}(|\eta'|^2 - g_0^2)$$

and it is easy to compute

$$\kappa_0 = -\frac{1}{8\nu}g_0^4 \quad \text{for} \quad n=3 \quad \text{and} \quad \kappa_0 = -\frac{1}{3\nu}g_0^3 \quad \text{for} \quad n=2. \tag{104}$$

The problem (102) has a solution $(U_0', \eta_n^{1-\gamma}Q_0)$ if and only if the right-hand side G_0 satisfies the compatibility condition

$$\int_\omega G_0\, d\eta' = 0. \tag{105}$$

From (105), taking into account (103), (104), we get

$$\begin{aligned} -\eta_n^{1-\gamma}\partial_n(\eta_n^{2(1-\gamma)}\partial_n q_0(\eta_n)) \int_\omega \Phi(\eta')\, d\eta' - & \\ -(\gamma-1)\eta_n^{3(1-\gamma)-1}\partial_n q_0(\eta_n) \int_\omega \eta' \cdot \nabla'\Phi\, d\eta' & = 0. \end{aligned}$$

Since

$$\int_\omega \eta' \cdot \nabla'\Phi(\eta')\, d\eta' = -(n-1)\int_\omega \Phi(\eta')\, d\eta' = -(n-1)\kappa_0,$$

the last relation yields

$$-\eta_n^{1-\gamma}\partial_n(\eta_n^{2(1-\gamma)}\partial_n q_0(\eta_n)) + (n-1)(\gamma-1)\eta_n^{3(1-\gamma)-1}\partial_n q_0(\eta_n) = 0. \tag{106}$$

Thus, the function $q_0(\eta_n)$ is not arbitrary; it satisfies the second order ordinary differential equation (106). Multiplying (106) by $\eta_n^{(n-2)(1-\gamma)}$, we rewrite it in the form

$$-\partial_n(\eta_n^{(n+1)(1-\gamma)}\partial_n q_0(\eta_n)) = 0. \tag{107}$$

Solving (107), we find

$$q_0(\eta_n) = \begin{cases} \mu_1\eta_n^{-(n+1)(1-\gamma)+1} + \mu_2, & \gamma \neq n(n+1)^{-1}, \\ \mu_1 \ln \eta_n + \mu_2, & \gamma = n(n+1)^{-1}. \end{cases} \tag{108}$$

Now, because of (108), (100)

$$U_{n0}(\eta', \eta_n) = \eta_n^{-(n-1)(1-\gamma)}\Phi(\eta') \begin{cases} \mu_1(n+1)(\gamma-1)+1, & \gamma \neq n(n+1)^{-1}, \\ \mu_1, & \gamma = n(n+1)^{-1}, \end{cases} \tag{109}$$

and the function G_0 takes the form

$$G_0(\eta', \eta_n) = -\mu_1 \eta_n^{-(n-2)(1-\gamma)-1} \Upsilon(\eta', \nabla')\Phi(\eta') \times$$
$$\times \begin{cases} (n+1)(\gamma-1)+1, & \gamma \neq n(n+1)^{-1}, \\ \\ 1, & \gamma = n(n+1)^{-1}, \end{cases}$$

where the operator $\Upsilon(\eta', \nabla')$ is given by

$$\Upsilon(\eta', \nabla') = (n-1)(\gamma-1) + (\gamma-1)\eta' \cdot \nabla'.$$

Comparing the power exponents of η_n in (109), (102), we conclude that the functions $U_0'(\eta', \eta_n)$ and $Q_0(\eta', \eta_n)$ can be taken in the form

$$U_0'(\eta', \eta_n) = \eta_n^{-(n-2)(1-\gamma)-1} U_0'(\eta'), \quad Q_0(\eta', \eta_n) = \eta_n^{-(n-1)(1-\gamma)-1} Q_0(\eta') \tag{110}$$

and we rewrite (102) as follows

$$\begin{cases} -\nu \Delta' U_0' + \nabla' Q_0 = 0 & \text{in } \omega, \\ \\ \operatorname{div}' U_0' = G_0 & \text{in } \omega, \\ \\ U_0' = 0 & \text{on } \partial\omega, \end{cases} \tag{111}$$

where

$$G_0(\eta') = -\mu_1 \Upsilon(\eta', \nabla')\Phi(\eta') \begin{cases} (n+1)(\gamma-1)+1, & \gamma \neq n(n+1)^{-1}, \\ \\ 1, & \gamma = n(n+1)^{-1}. \end{cases} \tag{112}$$

It is a well-known fact that in a bounded domain ω with smooth boundary $\partial\omega$ solutions of the Poisson equation (101) and of the Stokes system (111) are infinitely often differentiable up to the boundary and obey the estimates

$$|\partial_j^k \Phi(\eta')| \leq C_k, \tag{113}$$
$$|\partial_j^k U_0'(\eta')| + |\partial_j^k Q_0(\eta')| \leq C_k \, |\mu_1|. \tag{114}$$

for $j = 1, \ldots, n-1, \quad k = 0, 1, \ldots$. Therefore, we obtain (see (109), (110))

$$\begin{aligned} |\partial_j^k \partial_n^l U_{n0}(\eta', \eta_n)| &\leq c_{k,l} |\mu_1| \eta_n^{-(n-1)(1-\gamma)-l}, & k, l = 0, 1, \ldots, \\ |\partial_j^k \partial_n^l U_0'(\eta', \eta_n)| &\leq c_{k,l} |\mu_1| \eta_n^{-(n-2)(1-\gamma)-1-l}, & k, l = 0, 1, \ldots, \\ |\partial_j^k \partial_n^l Q_0(\eta', \eta_n)| &\leq c_{k,l} |\mu_1| \eta_n^{-(n+1)(1-\gamma)+1-l}, & k = 0, \ldots, l = 1, \ldots. \end{aligned} \tag{115}$$

Moreover, simple computations imply

$$\int_\omega U_{n0}(\eta', \eta_n)\, d\eta' = \mu_1 \kappa_0 \begin{cases} \frac{(n+1)(\gamma-1)+1}{\eta_n^{(n-1)(1-\gamma)}}, & \gamma \neq n(n+1)^{-1}, \\ \\ \eta_n^{-(n-1)(n+1)^{-1}}, & \gamma = n(n+1)^{-1}. \end{cases} \tag{116}$$

Let us define

$$\begin{aligned} u_0(x) &= U_0(x'x_n^{\gamma-1}, x_n), \\ p_0(x) &= P_0(x'x_n^{\gamma-1}, x_n) = q_0(x_n) + Q_0(x'x_n^{\gamma-1}, x_n). \end{aligned} \tag{117}$$

By the construction

$$\mathrm{div} u_0(x) = 0 \quad \text{in } \Omega_i, \qquad u_0(x) = 0 \quad \text{on } \partial\Omega_i \setminus \sigma_i(0)$$

and

$$\int_{\sigma_i} u_0 \cdot n\, dx' = \mu_1 \kappa_0 \begin{cases} (n+1)(\gamma-1)+1, & \gamma \neq n(n+1)^{-1}, \\ 1, & \gamma = n(n+1)^{-1}. \end{cases}$$

Thus, taking

$$\mu_1 = \begin{cases} F_i((n+1)(\gamma-1)+1)^{-1}\kappa_0^{-1}, & \gamma \neq n(n+1)^{-1}, \\ F_i \kappa_0^{-1}, & \gamma = n(n+1)^{-1}, \end{cases} \tag{118}$$

we have

$$\int_{\sigma_i} u_0 \cdot n\, dx' = F_i. \tag{119}$$

Furthermore, direct computations show that $u^{(0)}$, $p^{(0)}$ satisfy the Stokes system

$$\begin{aligned} -\nu\Delta u_0 + \nabla p_0 &= H_0 && \text{in } \Omega_i, \\ \mathrm{div} u_0 &= 0 && \text{in } \Omega_i, \\ u_0 &= 0 && \text{on } \partial\Omega_i \setminus \sigma_i(0) \end{aligned} \tag{120}$$

with the right-hand side H_0, subjected to the estimates

$$\begin{aligned} |D^\alpha H_{j,0}(x)| &\leq c\,|F_i| x_n^{-(n+1+|\alpha|)(1-\gamma)-(3+\alpha_n)\gamma}, \qquad j = 1, \ldots, n-1, \\ |D^\alpha H_{n,0}(x)| &\leq c\,|F_i| x_n^{-(n+1+|\alpha|)(1-\gamma)-(2+\alpha_n)\gamma}, \end{aligned} \tag{121}$$

where $\alpha = (\alpha_1, \cdots, \alpha_n)$, $|\alpha| = \alpha_1 + \cdots + \alpha_n$. The functions u_0, p_0 themselves obey the inequalities

$$\begin{aligned} |D^\alpha u_0'(x)| &\leq c\,|F_i| x_n^{-(n-1+|\alpha|)(1-\gamma)-\gamma}, \qquad |\alpha| \geq 0, \\ |D^\alpha u_{n,0}(x)| &\leq c\,|F_i| x_n^{-(n-1+|\alpha|)(1-\gamma)}, \qquad |\alpha| \geq 0, \\ |D^\alpha p_0(x)| &\leq c\,|F_i| x_n^{-(n+|\alpha|)(1-\gamma)}, \qquad |\alpha| \geq 1, \end{aligned} \tag{122}$$

$$|p_0(x)| \leq c \begin{cases} |F_i| x_n^{-(n+1)(1-\gamma)+1} + c_1, & \gamma \neq n(n+1)^{-1}, \\ |F_i| \ln x_n + c_1, & \gamma = n(n+1)^{-1}. \end{cases} \tag{123}$$

Remark. Inequality $(122)_2$ coincides with (94), where we take $g_i(x_n) = g_0 x_n^{1-\gamma}$, while $(122)_1$ states better decay at infinity for the components $u_{j,0}(x)$, $j = 1, \cdots, n-1$, of the velocity field u; we have in $(122)_1$ an additional vanishing factor $x_n^{-\gamma}$. The discrepancy H_0 also has at infinity better decay as ΔA. In $(121)_1$ we have an additional vanishing factor $x_n^{-\gamma(3+\alpha_n)}$ and in $(121)_2$ we have the factor $x_n^{-\gamma(2+\alpha_n)}$. This is the case, since we have already compensated the principal terms at infinity in equations (1).

Remark. Equation (107) describing q_0 is similar to the Reynolds equation, which is well-known in the theory of lubrication (see the lists of references in [32], [35]).

Remark. In Example 1 it was shown that divergence free vector fields with finite Dirichlet integral may have nonzero fluxes over the sections σ_i of the outlet to infinity Ω_i, having the form (31), if and only if there holds the condition

$$\int^{\infty} g_i^{-(n-1)}(t)\, dt < \infty$$

(see (36)). In the case where $g_i(t) = g_0 t^{1-\gamma}$ this yields

$$\int^{\infty} t^{-(n+1)(1-\gamma)}\, dt < \infty.$$

One has

$$\int^{\infty} t^{-(n+1)(1-\gamma)}\, dt = \infty, \quad \text{if} \quad \gamma < n(n+1)^{-1}$$

and

$$\int^{\infty} t^{-(n+1)(1-\gamma)}\, dt < \infty, \quad \text{if} \quad \gamma > n(n+1)^{-1}.$$

In the limit case $\gamma = n(n+1)^{-1}$ the logarithmic term appears in the asymptotic representation for the pressure function P.

The described procedure can be extended and one can compute the higher terms in the asymptotic expansion. Moreover, similar results can also be obtained in the case where the right-hand side f has a special series representation and for the nonlinear Navier-Stokes problem.

In order to justify the formal asymptotics, one has to estimate the remainder $u - u^{[N]}$. For the linear Stokes problem this can be done by applying the results of Section 3.2 and by using the weighted function spaces defined there. In the case of the nonlinear Navier-Stokes problem one can use Theorem 22. Because of the conditions (46), (47), it is possible to get the corresponding results for arbitrary large data only if $n = 3$ and if there holds the restriction

$$1/4 < \gamma_i < 1, \quad i = 1, \ldots, M. \tag{124}$$

If either $n = 3$ and (124) is violated or $n = 2$, the obtained asymptotics are justified only for small data. The detailed proofs of the listed results can be found in [36].

3.6. Open problems

Problem 4. The decay properties for the solution of the Navier-Stokes problem (2), (4) were proved in the three-dimensional case for arbitrary fluxes under the conditions (46), (47) on functions g_i (see Theorem 22). Without these conditions the results are proved only for small data (see [50]). The same restrictions one meets justifying the asymptotics of the solution (see (124)). This is related to the estimate (49) which is proved assuming (46), (47). Therefore, to remove these restrictions one needs to solve Problem 2.

The analogous results for the two-dimensional case are known only for small fluxes. One can try to extend these results for the case of arbitrary fluxes, may be, assuming certain symmetry of the flow domain.

4. Strong solutions in domains with conical outlets and in the aperture domain

4.1. Function spaces

Let us consider the domain Ω with conical outlets to infinity, i.e. $g_i(t) = \gamma_i(1+t)$. In this case we have

$$\exp(\beta_i \int_0^t g_i(\tau)^{-1}\, d\tau) = c\,(1+t)^{\beta_i/\gamma_i}.$$

Therefore, the exponential weights do not contribute to the norms of the function spaces. Moreover,

$$g_i(x_3) \sim (1+|x|) \quad \text{in} \quad \Omega_i$$

and we can take the weight $(1+|x|)$ instead of $g_i(x_n)$ in the definitions of the norms.

Condition (33) in this case is evidently violated. Hence, the results of Section 3 cannot be directly applied to the Stokes problem in domains with conical outlets to infinity.[5] However, in this situation it is possible to use the well-known technique which was developed for general elliptic problems in the classical paper of V.A. Kondrat'ev (1967) [20] (see also V.G. Maz'ya, B.A. Plamenevskii (1975, 1977, 1978) [27], [28], [29] and S.A. Nazarov, B.A. Plamenevskii (1994) [40]) and nowadays can be regarded as mathematical "folklore". This section is based on the paper of W. Borchers, K. Pileckas (1993) [7], where the detailed realization of the Kondrat'ev–Maz'ya–Plamenevskii scheme is given for the aperture domain.

Let Ω be an aperture domain (3). It can be regarded as a domain with two conical outlets to infinity $\Omega_1 = \{x \in \Omega : \quad x_3 > 0, \quad |x| > k_0\}$ and $\Omega_2 = \{x \in \Omega : \quad x_3 < 0, \quad |x| > k_0\}$. They coincide for large $|x|$ with the half spaces $\mathbb{R}^3_+ = \{x \in \mathbb{R}^3 : x_3 > 0\}$ and $\mathbb{R}^3_- = \{x \in \mathbb{R}^3 : x_3 < 0\}$ which we consider as cones of opening π.

[5] We should mention that the results of Section 3 are still valid if $\beta_i = 0$ and $æ_i$ are sufficiently small. However, this is not enough for our purposes.

Let us first specify the function spaces in the aperture domain Ω and in the half space $I\!R^3_+$. For nonnegative integer l, $1 < s < \infty$ and $-\infty < æ < \infty$ the space $V^{l,s}_æ(I\!R^3_+)$ is defined as a completion of $C^\infty_0(I\!R^3_+)$ in the norm

$$\|u; V^{l,s}_æ(I\!R^3_+)\| = (\sum_{|\alpha|=0}^{l} \int_{I\!R^3_+} |x|^{s(æ+|\alpha|-l)} |D^\alpha u|^s dx)^{1/s}.$$

As usual $L^s_æ(I\!R^3_+) = V^{0,s}_æ(I\!R^3_+)$ and $V^{-1,s}_æ(I\!R^3_+)$ denotes the dual space to $V^{1,s'}_{-æ}(I\!R^3_+)$. The weighted Hölder space $C^{l,\delta}_æ(I\!R^3_+), l \geq 0, 0 < \delta < 0, -\infty < æ < \infty$, consists of functions u, continuously differentiable up to the order l in $I\!R^3_+$, for which the norm

$$\|u; C^{l,\delta}_æ(I\!R^3_+)\| = \sum_{|\alpha|\leq l} \sup |x|^{æ-l-\delta+|\alpha|} |D^\alpha u(x)| + \sum_{|\alpha|=l} \sup |x|^æ [D^\alpha u]_\delta(x)$$

is finite. Here, the supremum is taken over $x \in I\!R^3_+$ and

$$[u]_\delta(x) = \sup_{\substack{0<|x-y|<|x|/2 \\ y\in I\!R^3_+}} \frac{|u(x)-u(y)|}{|x-y|^\delta}.$$

In the aperture domain Ω we use the space $L^2_æ(\Omega)$, consisting of functions with the finite norm

$$\|u; L^q_æ(\Omega)\| = \|u; L^q(\Omega_{(k_0+1)})\| + \|\zeta u; L^q_æ(I\!R^3_+)\| + \|\zeta u; L^q_æ(I\!R^3_-)\|$$

and the space $V^{1,2}_æ(\Omega)$, which is the completion of $C^\infty_0(\Omega)$ in the norm

$$\|u; V^{1,q}_æ(\Omega)\| = \|u; W^{1,q}(\Omega_{(k_0+1)})\| + \|\zeta u; V^{1,q}_æ(I\!R^3_+)\| + \|\zeta u; V^{1,q}_æ(I\!R^3_-)\|.$$

Here ζ is a smooth cut–off function with $\zeta(|x|) = 1$ for $|x| > k_0 + 1$ and $\zeta(|x|) = 0$ for $|x| < k_0$. $V^{-1,2}_æ(\Omega)$ is the dual space to $V^{1,2}_{-æ}(\Omega)$.

4.2. Stokes problem in the aperture domain

We consider the Stokes system (1) supplemented either by the additional flux condition

$$\int_S u_3(x', x_3) dx' = F \tag{125}$$

with prescribed $F \in I\!R^1$ or by the pressure drop condition

$$p_* = \lim_{\substack{|x|\to\infty \\ x_3>0}} p(x) - \lim_{\substack{|x|\to\infty \\ x_3<0}} p(x) \tag{126}$$

with given $p_* \in I\!R^1$. According to the methods of [20], we first study the Stokes problem in a half space $I\!R^3_+$:

$$\begin{aligned} -\nu\Delta u + \nabla p &= f \quad \text{in} \quad I\!R^3_+, && (127)\\ \mathrm{div} u &= g \quad \text{in} \quad I\!R^3_+, && (128)\\ u &= 0 \quad \text{on} \quad \{x \in I\!R^3 : \quad x_3 = 0\}. && (129) \end{aligned}$$

The main existence and uniqueness theorems for the Stokes problem in weighted Hilbert spaces, as well as the asymptotic behaviour of solutions, are based on a priori estimates for the corresponding linear operator with a complex parameter on the unit hemisphere. This operator is obtained from the Stokes problem in $I\!R^3_+$ by applying the Mellin transform in the radial direction:

$$\hat{u}(\lambda, \sigma) = \int_0^\infty r^{\lambda-1} u(\sigma r) dr, \quad \sigma = x/ \mid x \mid . \qquad (130)$$

We denote by Γ the upper part of the unit sphere, i.e. the hemisphere which is contained in $I\!R^3_+$. It is well-known that the operator $Mu := \hat{u}$ extends to a bounded and invertible operator

$$\mathcal{M} : L^2_{æ}(I\!R^3_+) \to L^2(æ + 3/2 - i\infty, æ + 3/2 + i\infty; L^2(\Gamma)). \qquad (131)$$

In order to apply $\mathcal{M}$ to the Stokes system (127)-(129), we first rewrite it in spherical coordinates

$$x = r\phi(\theta, \varphi), \quad \phi(\theta, \varphi) = (\sin\theta\cos\varphi, \ \sin\theta\sin\varphi, \ \cos\theta), \qquad (132)$$

with $r = |x|$, $0 \le \theta < \pi$, $0 \le \varphi < 2\pi$. For the spherical components $u_\phi = (u^r, u^\theta, u^\varphi)$ of u defined by

$$u_\phi = u^r e_r + u^\theta e_\theta + u^\varphi e_\varphi, \qquad (133)$$

where $e_r = \phi$, $e_\theta = \partial_\theta \phi$, $e_\varphi = \sin^{-1}\theta \partial_\varphi \phi$ are unit tangent vectors to the coordinate lines, we obtain the following boundary value problem on Γ:

$$\begin{aligned} \nu\Big((-\lambda(\lambda-1)+2)u^r - \Delta_\Gamma u^r + 2\mathrm{div}_\Gamma(u^\theta, u^\varphi)\Big) - (\lambda+1)p &= f^r,\\ \nu(-\lambda(\lambda-1)u^\theta - \Delta_\Gamma u^\theta) - 2\nu\left(\partial_\theta u^r - \frac{u^\theta}{2\sin^2\theta} - \frac{\cos\theta}{\sin^2\theta}\partial_\varphi u^\varphi\right) + \partial_\theta p &= f^\theta,\\ \nu(-\lambda(\lambda-1)u^\varphi - \Delta_\Gamma u^\varphi) - \frac{2\nu}{\sin\theta}\left(\partial_\varphi u^r + \frac{\cos\theta}{\sin\theta}\partial_\varphi u^\theta - \frac{u^\varphi}{2\sin\theta}\right) + \frac{1}{\sin\theta}\partial_\varphi p &= f^\varphi,\\ (\lambda-2)u^r - \mathrm{div}_\Gamma(u^\theta, u^\varphi) &= g,\\ u^r(\pi/2, \varphi) = u^\theta(\pi/2, \varphi) = u^\varphi(\pi/2, \varphi) &= 0. \end{aligned}$$

and denote by $\mathcal{L}(\lambda)$ the corresponding operator.

Theorem 24 *The eigenvalues*[6] *of* $\mathcal{L}(\lambda)$ *are positive or negative integers. The number* $\lambda = 2$ *is the smallest nonnegative eigenvalue of* $\mathcal{L}(\lambda)$. *Moreover,* $\lambda = 2$ *is a simple eigenvalue. The corresponding eigenspace is spanned by*

$$\begin{aligned} E_1 = (b_1, p_1) &= (x_3^2 x,\ 2\nu + 2\nu x_1^2) \\ E_2 = (b_2, p_2) &= (x_3 x_1 x,\ 2\nu x_3 x_1) \\ E_3 = (b_3, p_3) &= (x_3 x_2 x,\ 2\nu x_3 x_2) \end{aligned} \tag{134}$$

with $x \in \Gamma$. *The adjoint eigenspace (eigenvalue* $\lambda = -1$*) is spanned by*

$$\begin{aligned} E_1^* = (b_1^*, p_1^*) &= (0,\ 1) \\ E_2^* = (b_2^*, p_2^*) &= (x_3 e_1,\ 0) \\ E_3^* = (b_3^*, p_3^*) &= (x_3 e_2,\ 0), \end{aligned} \tag{135}$$

where $e_i, i = 1, 2, 3$ *are the cartesian unit vectors.*

The information about the spectrum of the operator $\mathcal{L}(\lambda)$ allows us to prove the following result.

Theorem 25 (i) *Let* $f \in V_{æ}^{-1,2}(\mathbb{R}_+^3)$, $g \in L_{æ}^2(\mathbb{R}_+^3)$, *with* $æ \neq$ *integer* $+ 1/2$ *for* $|æ| \geq 1$. *Then there exists a unique weak solution* $(u, p) \in V_{æ}^{1,2}(\mathbb{R}_+^3) \times L_{æ}^2(\mathbb{R}_+^3)$ *of Stokes problem* (127) – (129). *Moreover, the following estimate is valid*

$$\|u; V_{æ}^{1,2}(\mathbb{R}_+^3)\| + \|p; L_{æ}^2(\mathbb{R}_+^3)\| \leq c\,(\|f; V_{æ}^{-1,2}(\mathbb{R}_+^3)\| + \|g; L_{æ}^2(\mathbb{R}_+^3)\|) \tag{136}$$

with a constant $c > 0$ *independent of* f *and* g.

(ii) *Let* $f \in V_{æ_2}^{-1,2}(\mathbb{R}_+^3) \cap V_{æ_1}^{-1,2}(\mathbb{R}_+^3), g \in L_{æ_2}^2(\mathbb{R}_+^3) \cap L_{æ_1}^2(\mathbb{R}_+^3)$ *with* $5/2 > æ_2 > æ_1 > -5/2$, $æ_i \neq \pm 3/2$, $i = 1, 2$. *Then the solutions* $(u^1, p^1) \in V_{æ_1}^{1,2}(\mathbb{R}_+^3) \times L_{æ_1}^2(\mathbb{R}_+^3)$ *and* $(u^2, p^2) \in V_{æ_2}^{1,2}(\mathbb{R}_+^3) \times L_{æ_2}^2(\mathbb{R}_+^3)$ *of problem* (127) – (129) *are related by*

$$u^1 = \sum_j \sum_{k=1}^3 c_{jk} r^{-\lambda_j} U^{(j,k)} + u^2, \tag{137}$$

$$p^1 = \sum_j \sum_{k=1}^3 c_{jk} r^{-\lambda_j - 1} P^{(j,k)} + p^2, \tag{138}$$

where c_{jk} *are constants,* $(U^{(j,k)}, P^{(j,k)})$ *are eigenvectors corresponding to eigenvalues* λ_j *of the operator* $\mathcal{L}(\lambda)$ *and the sums over* j *in* (137) *and* (138) *are taken over all eigenvalues* λ_j, *lying in the strip* $æ_1 + 1/2 < Re\lambda < æ_2 + 1/2$.

[6] As usual, the poles of the Green function of the problem are called eigenvalues. The corresponding eigenvectors are (by definition) solutions of the corresponding homogeneous problem.

Remark. By Theorem 24 there can be at most two eigenvalues, $\lambda_1 = -1$ and $\lambda_2 = 2$, in the strip $æ_1 + 1/2 < \operatorname{Re}\lambda < æ_2 + 1/2$. Therefore, $U^{(j,k)} = b_k^j$, with $b_k^1 = b_k^*$ and $b_k^2 = b_k$ defined by (134) and (135); $P^{(j,k)}$ are the corresponding pressure functions.

Now we apply the foregoing results to study the asymptotic behaviour of a weak solution to Stokes problem (1), (125) (or (1), (126)) in the aperture domain. In what follows, we derive precise information on the stronger decay of solutions, assuming a stronger decay of the volume forces f as $r = |x|$ tends to infinity.

Denote by $(U_\pm^{(k)}, P_\pm^{(k)})$ the restrictions of functions $\pm E_k(\pm x/|x|)$, defined in Theorem 24, to the half spaces $\mathbb{R}_\pm^3$. Then the pairs

$$(r^{-2}U_\pm^{(k)}, r^{-3}P_\pm^{(k)}), \qquad k = 1, 2, 3 \tag{139}$$

define singular solutions of the homogeneous Stokes problem in the half spaces $\mathbb{R}_\pm^3$.

Theorem 26 (i) *Let $F \in \mathbb{R}^1, f \in V_æ^{-1,2}(\Omega)$ with $0 \le æ < 3/2$. Then there exists a unique weak solution $u_F \in V_æ^{1,2}(\Omega)$ of problem (1), (125). The corresponding pressure function $p_F \in L^2_{loc}(\overline{\Omega})$ is unique modulo a constant and satisfies the representation*

$$\zeta p_F(x) = p^\pm + q(x), \;\; x \in \mathbb{R}_\pm^3, \tag{140}$$

where $p^\pm$ are constants and $q(x) \in L^2_æ(\mathbb{R}_\pm^3)$.

(ii) *Let $p_* \in \mathbb{R}^1, f \in V_æ^{-1,2}(\Omega)$ with $0 \le æ < 3/2$. Then problem (1), (126) has a unique weak solution $u_d \in V_æ^{1,2}(\Omega)$. The corresponding pressure function $p_d \in L^2_{loc}(\overline{\Omega})$ is unique modulo a constant and satisfies (126) with $p^+ - p^- = p_*$.*

(iii) *Let $f \in V_æ^{-1,2}(\Omega)$ with $3/2 < æ < 5/2$. Then the weak solutions (u_F, p_F) and (u_d, p_d) of problems (1), (125) and (1), (126) admit the representations*

$$\zeta u = \sum_{k=1}^{3} C_\pm^{(k)} r^{-2} U_\pm^{(k)} + v, \quad x \in \mathbb{R}_\pm^3, \tag{141}$$

$$\zeta p = p^\pm + \sum_{k=1}^{3} C_\pm^{(k)} r^{-3} P_\pm^{(k)} + q, \quad x \in \mathbb{R}_\pm^3, \tag{142}$$

with $v \in V_æ^{1,2}(\mathbb{R}_\pm^3), q \in L^2_æ(\mathbb{R}_\pm^3)$. Here u, p stands either for u_F, p_F or for u_d, p_d. For the solution u_F, p_F of problem (1), (125) the constants $C_\pm^{(1)}$ are given by $C_+^{(1)} = -C_-^{(1)} = 3F/2\pi$ and for the solution u_d, p_d of problem (1), (126) there holds the equality $p^+ - p^- = p_*$.

Theorem 27 *Let the right-hand side $f \in H^*(\Omega)$ and let (u, p) be the corresponding weak solution either of (1), (125) or of (1), (126) (see Theorem 14).*

(i) *Assume that* $\zeta f \in V^{l,s}_{æ}(I\!R^3_\pm)$ *with* $l \geq -1$ *being an integer ,* $s > 1$ *,* $0 \leq æ - l - 2 + 3/s < 2$. *Then* $\zeta u \in V^{l+2,s}_{æ}(I\!R^3_\pm)$ *and* ζp *satisfies the representation* (142) *with* $q \in V^{l+1,s}_{æ}(I\!R^3_\pm)$.

(ii) *If* $\zeta f \in V^{l,s}_{æ}(I\!R^3_\pm)$ *with* $2 < æ - l - 2 + 3/s < 3$, *then* $\zeta u, \zeta p$ *satisfy* (141), (142) *with* $v \in V^{l+2,s}_{æ}(I\!R^3_\pm)$ *and* $q \in V^{l+1,s}_{æ}(I\!R^3_\pm)$.

(iii) *Let* $\zeta f \in C^{l,\delta}_{l+2+\delta+æ}(I\!R^3_\pm)$, $l \geq 0$ *being an integer,* $\delta \in (0,1)$, $2 \leq æ \leq 3$. *Then* $\zeta u, \zeta p$ *satisfy* (141), (142) *with* $v \in C^{l+2,\delta}_{l+2+\delta+æ}(I\!R^3_\pm)$, $q \in C^{l+1,\delta}_{l+2+\delta+æ}(I\!R^3_\pm)$.

The proof of this theorem is based on Theorem 26.

Theorem 28 *Let* $u_h \in \widehat{H}(\Omega), p_h \in L^2_{loc}(\Omega)$ *be a weak solution of the homogeneous* ($f = 0$) *Stokes problem* (1), (125) *satisfying condition* (125) *with* $F = 1$. *Then* u_h, p_h *admit the asymptotic representation as* $|x| \to \infty$

$$
\begin{aligned}
u_h &= \pm\frac{3}{2\pi} r^{-2} U^{(1)}_\pm + \sum_{k=2}^{3} C^{(k)}_{\pm,h} r^{-2} U^{(k)}_\pm + O(r^{-3+\varepsilon}), \\
p_h &= p^\pm_h \pm \frac{3}{2\pi} r^{-3} P^{(1)}_\pm + \sum_{k=2}^{3} C^{(k)}_{\pm,h} r^{-3} P^{(k)}_\pm + O(r^{-4+\varepsilon}),
\end{aligned}
\qquad (143)
$$

$x \in I\!R^3_\pm$, $\varepsilon > 0$. *Formulas* (143) *allow differentiation.*

This theorem is a direct consequence of Theorem 27 (iii) with $f = 0$ and $æ = 3 - \varepsilon$.

In the subsequent theorem we find the relation between the flux F and the pressure drop p_*.

Theorem 29 *Let* $f \in H^*(\Omega)$ *and* (u,p) *be a weak solution either of problem* (1), (125) *or of problem* (1), (126). *Denote respectively by* F *and* p_* *the flux and the pressure drop, corresponding to this solution. (For the problem* (1), (125) F *is fixed and* p_* *is unknown, and for the problem* (1), (126) p_* *is fixed,* F *is unknown.) Then there holds*

$$
\int_\Omega f \cdot u_h \, dx = p_* - F p_{*h}, \qquad (144)
$$

where (u_h, p_h) *is the solution of the homogeneous Stokes problem* (1), (125) *with* $F = 1$ *and* $p_{*h} = p^+_h - p^-_h$.

4.3. Navier–Stokes problem in the aperture domain

We consider the steady Navier–Stokes system (2) supplement either by the additional flux condition (125) or by the pressure drop condition (126). There holds

Theorem 30 *Assume that* $f \in V_{æ}^{-1,s}(\Omega)$ *with* $3/2 < æ < 5/2$ *and let* $u \in V_{1/2+\varepsilon}^{1,2}(\Omega)$, $\varepsilon > 0$, *be a solution either of problem* (2) *with additional flux condition* (125) *or with pressure drop condition* (126). *Then* u *and the corresponding pressure function* p *admit the asymptotic representation* (141), (142) *with* $v \in V_{æ}^{1,2}(\mathbb{R}^3_{\pm}), q \in L^2_{\beta}(\mathbb{R}^3_{\pm})$.

The proof of the theorem is based on the results for the linear Stokes problem and on bootstrap arguments.

The next theorem shows, that there exist solutions satisfying the requirements of the previous theorem, at least for small data (the proof of it is based on the Banach contraction principle).

Theorem 31 (i) *Let* $|F|$ *and* $\|f; V_{æ}^{-1,2}(\Omega)\|$ *be sufficiently small,* $1/2 < æ < 3/2$. *Then there exists a unique solution* (u, p) *of problem* (2), (125) *with* $u \in V_{æ}^{1,2}(\Omega)$.

(ii) *For sufficiently small* $|p_*|$ *and* $\|f; V_{æ}^{-1,2}(\Omega)\|$, $1/2 < æ < 3/2$, *there exists a unique solution* (u, p) *of problem* (2), (126) *with* $u \in V_{æ}^{1,2}(\Omega)$.

(iii) *If* $3/2 < æ < 5/2$, *then* (u, p) *satisfies the asymptotic representations* (141), (142) *with* $v \in V_{æ}^{1,2}(\mathbb{R}^3_{\pm})$, $q \in L^2_{æ}(\mathbb{R}^3_{\pm})$.

The pointwise decay of the solution is proved in the next theorem. For simplicity we assume $f = 0$.

Theorem 32 *Let* $u \in V_{1/2+\varepsilon}^{1,2}(\Omega)$, $\varepsilon > 0$, *be a solution either to problem* (2), (125) *or to problem* (2), (126) *and let* $f = 0$. *Then* u *and the corresponding pressure function* p *satisfy the following asymptotic relations for* $x \in \mathbb{R}^3_{\pm}$, $|x| \to \infty$

$$\begin{aligned} u(x) &\sim \sum_{k=1}^{3} c_{\pm}^{(k)} r^{-2} U_{\pm}^{(k)} + O(r^{-3+\delta}), \\ p(x) &\sim p^{\pm} + \sum_{k=1}^{3} c_{\pm}^{(k)} r^{-3} P_{\pm}^{(k)} + O(r^{-4+\delta}). \end{aligned} \tag{145}$$

In (145) $\delta > 0$, $U_{\pm}^{(k)}$ *and* $P_{\pm}^{(k)}$ *are the functions defined in* (139). *In the case of problem* (2), (125) *the constants* $c_{\pm}^{(1)}$ *are defined by* $c_{+}^{(1)} = -c_{-}^{(1)} = 3F/2\pi$ *and in the case of problem* (2), (126) *there holds the equality* $p^+ - p^- = p_*$. *The formulas* (145) *can be differentiated.*

Remark. For example in $\mathbb{R}^3_+$ we obtain, inserting the explicit expressions for $U_{\pm}^{(k)}, P_{\pm}^{(k)}$, the following

$$\begin{aligned} u(x) &= \frac{3F}{2\pi}\frac{x_3^2 x}{|x|^5} + c_2^+ \frac{x_3 x_1 x}{|x|^5} + c_3^+ \frac{x_3 x_2 x}{|x|^5} + O(r^{-3+\delta}), \\ p(x) &= p^+ + \frac{3F}{2\pi}\left(\frac{2\nu}{|x|^3} + \frac{2\nu x_3^2}{|x|^5}\right) + c_2^+ \frac{2\nu x_3 x_1}{|x|^5} + c_3^+ \frac{2\nu x_3 x_2}{|x|^5} + O(r^{-4+\delta}). \end{aligned}$$

This implies in particular, that in each plane $x_3 = const.$ the terms which carry the flux are decaying more rapidly than the other terms. Similarly it is observed, that the velocity component parallel to the x_3-axis decays more rapidly. Moreover, in every domain

$$\{x = (x_1, x_2, x_3) : |x_3| < C(x_1^2 + x_2^2)^{\alpha/2}\}, \ \ 0 \leq \alpha < 1, \ C > 0,$$

the velocity decays like $O(r^{-3+\alpha})$ displaying a "wide" jet region.

Let us prove now the uniqueness for small data of a weak solution $u \in \widehat{H}(\Omega)$ to the Navier-Stokes problem, satisfying the energy inequality

$$\int_\Omega \nabla u \, : \, \nabla u \, dx \leq \langle f, u \rangle - p_* F. \tag{146}$$

Here and it what follows $\langle *, * \rangle$ is the duality pairing. The idea of the proof is essentially to compare the weak solution satisfying (146) with a "strong" solution from Theorem 31, which decays at large distances sufficiently fast, by means of a "mixed" energy method involving integrals which always contain the product of both solutions. Note that the weak solutions of Theorem 15 satisfy the energy inequality (146).

The next theorem states the energy equality for the strong solutions of Theorem 31.

Theorem 33 *Let* $v \in \widehat{H}(\Omega) \cap V^{1,2}_{æ}(\Omega)$, $1/2 < æ < 3/2$, *be a "strong" solution either of problem* (2), (125) *or of problem* (2), (126), *according to Theorem* 31. *Then* v *satisfies the energy equality*

$$\int_\Omega \nabla v \, : \, \nabla v \, dx = \langle f, v \rangle - q_* Q, \tag{147}$$

where

$$\begin{aligned} q_* &= q^+ - q_-, \quad q^\pm = \lim_{|x| \to \infty, \ x \in I\!R^3_\pm} q(x), \\ Q &= \int_S v_3(x', x_3) \, dx. \end{aligned}$$

In the case of problem (2), (125) *we have* $Q = F$, *while in the case of problem* (2), (126) *it is* $q_* = p_*$.

We are now in a position to prove the following uniqueness result.

Theorem 34 *Let* $f \in V^{-1,2}_{1/2+\varepsilon}(\Omega)$ *for some* $\varepsilon > 0$. *If* $\|f; V^{-1,2}_{1/2+\varepsilon}(\Omega)\|$, $|F|$ *and* $|p_*|$ *are sufficiently small, then the solution* $u \in \widehat{H}(\Omega)$ *either of problem* (2), (125) *or of problem* (2), (126), *satisfying the energy inequality* (146), *is unique and coincides with the strong solution* $v \in V^{1,2}_{1/2+\varepsilon}(\Omega)$ *of the same problem.*

Proof. Since the proof is the same for both problems, we shall consider the case of problem (2), (125). In the integral identities for u and v we take $\varphi = v$ and $\eta = u$, respectively. Then, summing the resulting identities and taking into account that $\operatorname{div} u = \operatorname{div} v = 0$, we obtain

$$\begin{aligned} 2\nu \int_\Omega \nabla u : \nabla v \, dx &= \langle f, u\rangle + \langle f, v\rangle \\ &\quad - p_* F - q_* F - \int_\Omega ((u\cdot\nabla)u\cdot v + (v\cdot\nabla)v\cdot u)\, dx. \quad (148) \end{aligned}$$

From the identity

$$\int_\Omega \nabla(u-v) : \nabla(u-v)\, dx = \int_\Omega \nabla u : \nabla u \, dx + \int_\Omega \nabla v : \nabla v\, dx - 2\int_\Omega \nabla u : \nabla v \, dx$$

and from (146), (147) and (148) we obtain

$$\nu \int_\Omega \nabla(u-v) : \nabla(u-v)\, dx \le \int_\Omega ((u\cdot\nabla)u\cdot v + (v\cdot\nabla)v\cdot u))\, dx$$

$$= \int_\Omega (((u-v)\cdot\nabla)(u-v)\cdot v + ((u-v)\cdot\nabla)v\cdot v)dx + \int_\Omega ((v\cdot\nabla)v\cdot u + (v\cdot\nabla)u\cdot v)\, dx. \quad (149)$$

By weighted embedding theorems we have

$$\int_\Omega |u-v|^2(1+|x|)^2 dx + \int_\Omega |v|^4\, dx < \infty,$$

and, consequently, there is a sequence $\{R_s\}$ such that

$$\lim_{R_s\to\infty} R_s \int_{S^\pm_{R_s}} |u-v|^2 R_s^{-2} dS = \lim_{R_s\to\infty} R_s \int_{S^\pm_{R_s}} |v|^4 dS = 0.$$

Thus,

$$\begin{aligned} 2\left|\int_\Omega ((u-v) \cdot \nabla)v\cdot v)\, dx\right| &= \left|\lim_{R_s\to\infty}\int_{\Omega(R_s)} ((u-v)\cdot\nabla)v^2)dx\right| \\ &= \left|\lim_{R_s\to\infty}\int_{S^+_{R_s}\cup S^-_{R_s}} (u-v)\cdot n v^2\, dS\right| \\ &\le \lim_{R_s\to\infty}\left(R_s\int_{S^+_{R_s}\cup S^-_{R_s}} (u-v)^2 R_s^{-2} dS\right)^{1/2}\left(R_k\int_{S^+_{R_s}\cup S^-_{R_s}} |v|^4 dS\right)^{1/2} \\ &= 0 \end{aligned}$$

and

$$\begin{aligned} \int_\Omega (v\cdot\nabla)v\cdot u\, dx &= \lim_{R_s\to\infty}\int_{\Omega(R_s)} (v\cdot\nabla)v\cdot u\, dx \\ &= -\lim_{R_s\to\infty}\int_{\Omega(R_s)} (v\cdot\nabla)u\cdot v\, dx + \lim_{R_s\to\infty}\int_{S^+_{R_s}\cup S^-_{R_s}} (v\cdot n)(u\cdot n)dS \\ &= -\int_\Omega (v\cdot\nabla)u\cdot v\, dx. \end{aligned}$$

In virtue of these estimates the last three terms at the right–hand side of (149) are equal to zero and so we find

$$\nu \int_\Omega \nabla w : \nabla w \, dx \leq \int_\Omega (w \cdot \nabla) w \cdot v \, dx$$

with $w = u - v$. Estimating the integral at the right by the Hölder inequality and applying a weighted embedding theorem, we obtain

$$\begin{aligned} \nu \int_\Omega \nabla w : \nabla w \, dx &\leq \|\nabla(u-v)\|_{2,\Omega} \|w; L^4_{-1/4}(\Omega)\| \|v; L^4_{1/4}(\Omega)\| \\ &\leq c\|\nabla w\|^2_{2,\Omega} \|v; V^{1,2}_{1/2+\varepsilon}(\Omega)\| \leq c\|\nabla w\|^2_{2,\Omega}(|F| + \|f; V^{-1,1}_{1/2+\varepsilon}(\Omega)\|). \end{aligned}$$

Therefore, if

$$|F| + \|f; V^{-1,2}_{1/2+\varepsilon}(\Omega)\| < \nu/c,$$

we deduce

$$\nu \int_\Omega \nabla w : \nabla w \, dx = 0,$$

which concludes the proof of the theorem.

The presented uniqueness theorem was proved in [8].

4.4. Remarks on the two-dimensional case

The asymptotic properties of solutions of the Navier-Stokes problem in the two-dimensional aperture domain were investigated by S.A. Nazarov [31] and G.P. Galdi, M.R. Padula, V.A. Solonnikov [10]. It is proved in [31], [10] that in the two-dimensional case the solution decays at infinity as $O(|x|^{-1})$. However, this result is obtained for small data and only under additional symmetry assumptions on the flow domain and the data of the problem. The analogous result is also proved in a domain with two outlets which coincide with infinite corners

$$K_\pm = \{x \in I\!R^2 : r > R_0, \ |\varphi - \varphi_\pm| < \alpha_\pm\},$$

where (r, φ) are polar coordinates in $I\!R^2$. Moreover, if the angles $\alpha_\pm$ are less than $\pi/2$, the symmetry assumption is not necessary (see [31]).

4.5. Open problems

Problem 5. The results obtained for the three-dimensional aperture problem can be compared with those of R. Finn [13] and K.I. Babenko [5] for the uniform three-dimensional flow past an obstacle with nonzero constant velocity at infinity. In [13] it has been proved that each solution of this problem satisfying the relation $u(x) \sim O(r^{-1/2-\varepsilon})$ for large r, decays at infinity like r^{-1} and possesses a parabolic wake region. For small data the existence of such solutions has been proved in [13]. Lateron K.I. Babenko [5] has shown that each solution of this problem with finite Dirichlet integral has the same asymptotic behaviour.

In [7] it is proved that for the aperture problem every solution which decays in a certain sense as $O(r^{-1-\varepsilon})$ admits the asymptotic representation (145) (see Theorem 32 and the Remark after it). The existence of such solution for small data is proved in Theorem 31. However, for the aperture problem the result analogous to that of K.I. Babenko is not known, i.e. it is not known whether an arbitrary solution with a finite Dirichlet integral has the asymptotic representation (145).

Concerning the two-dimensional aperture problem the corresponding results are available only for small data and under symmetry assumptions. It would be interesting to remove one or both of these restrictions.

5. Final notes

5.1. Asymptotic properties of solutions in an infinite layer

In Example 5 we have presented an exact solution of the Stokes problem in a layer $\mathbf{L}$ (see the formula (59)). This solution has a nonzero flux and satisfies the homogeneous equations (1) everywhere in $\mathbf{L}$ except at the axis $|x'| = 0$. On this axis it has a singularity ($u = O(|x'|^{-1})$ as $|x'| \to 0$). Differentiating the formula (59) one can obtain a row of singular solutions to the Stokes system (1) in $\mathbf{L}$. These solutions are defined by ($k = 1, 2, \ldots$.)

$$\begin{aligned} u_i^{(k,\pm)}(r,\varphi,z) &= \frac{1}{2\nu} z(z-1)\frac{\partial p^{(k,\pm)}}{\partial x_i}(r,\varphi) \text{ for } i=1,2, \quad u_3^{(k,\pm)}(r,\varphi,z)=0, \\ p^{(k,+)}(r,\varphi) &= r^{-k}\cos(k\varphi), \quad p^{(k,-)}(r,\varphi) = r^{k}\sin(k\varphi) \end{aligned} \tag{150}$$

(remember that (r, φ, z) are cylindrical coordinates in $I\!R^3$). It is easy to verify that $u^{(k,\pm)}(r,\varphi,z)$ have zero fluxes and that

$$u^{(k,\pm)}(r,\varphi,z) = O(|x'|^{-k-1}).$$

Let $\zeta(r)$ be a smooth cut-off function with $\zeta(r) = 0$ if $r < d/2$ and $\zeta(r) = 1$ if $r > d$. Then the pair $(v^{(k,\pm)},\ q^{(k,\pm)}) = (\zeta u^{(k,\pm)},\ \zeta p^{(k,\pm)})$ is the solution of the following Stokes problem

$$\begin{aligned} -\nu\Delta v^{(k,\pm)} + \nabla q^{(k,\pm)} &= f^{(k,\pm)} \quad \text{in} \quad \Omega, \\ \operatorname{div} v^{(k,\pm)} &= g^{(k,\pm)} \quad \text{in} \quad \Omega, \\ v^{(k,\pm)} &= 0 \quad \text{on} \quad \partial\Omega. \end{aligned}$$

It is easy to see that the right-hand side $(f^{(k,\pm)},\ g^{(k,\pm)})$ has a compact support in the domain $\{x \in \mathbf{L} :\ d/2 < r < d\}$. Thus, we have constructed a sequence of solutions to the Stokes problem (1) with right-hand sides having compact supports. These solutions have zero fluxes (for sufficiently remote sections) and decay like $O(|x'|^{-k-1})$ as $|x'| \to \infty$, $k = 1, 2, \dots$. This example shows that the solutions in the layer-like domain do not decay exponentially fast (in contrast to the pipe or two-dimensional strip).

The solutions (150) have an anisotropic structure: the third component $u_3^{(k,\pm)}$ of the velocity field decays faster (as $r \to \infty$) than the first two components $u_1^{(k,\pm)}$ and $u_2^{(k,\pm)}$ (in fact, $u_3^{(k,\pm)} = 0$). Moreover, one can see by direct differentiation that the derivatives $\partial u^{(k,\pm)}/\partial x_i$, $i = 1, 2$, and $\partial u^{(k,\pm)}/\partial x_3$ also have different decay rate as $r \to \infty$.

In the forthcoming papers of S.A. Nazarov and the author we investigate the properties to the Stokes operator in a class of adequate weighted function spaces which reflect the anisotropic behaviour of exact solutions. Moreover, in these papers we construct and justify the asymptotics of solutions of the Stokes problem (1). We can also recommend the paper [37] where the analogous results were obtained for the Neumann problem for elliptic equations of second oder.

5.2. Asymptotic conditions at infinity

In this paper we have discussed different problems for Stokes and Navier-Stokes equations in infinite domains. However, it is self–understood, that there are no infinite volumes of liquid in reality and, hence, these problems should be considered as model problems. Exactly such problems are used by engineers while solving certain practical problems. Performing the computer simulation of the flow in long thin pipes, engineers, at first, replace them by semi–infinite cylinders and suggest some sensible conditions at infinity in each semi–cylinder based on the intuition, experiments, and engineering know–how. After this, the semi–infinite cylinders must be cut again, in order to apply numerical schemes for the flow simulation (e.g. [12]). In this connection the proper formulation of "boundary conditions" at infinity gain decisive significance for the adequate description of real situations.

All the results mentioned in this paper are related to two types of asymptotic conditions at infinity: either prescribed fluxes, or prescribed pressure drops. However, these conditions do not serve all possible physical phenomena which can occur in reality. For example, there appear the natural questions how to calculate the distribution of pressures in outlets to infinity, corresponding to the solution with prescribed fluxes, and vice versa, how to calculate the distribution of fluxes for the solution with prescribed pressure drops. Another natural problem is, for instance, the following: how to calculate the fluxes in a system of pipes if the liquid flows in through one of the pipes (say $\mathbf{\Pi}_1$) and then distributes over the rest pipes $\mathbf{\Pi}_2, \dots, \mathbf{\Pi}_M$, i.e. the flux is known only in $\mathbf{\Pi}_1$. From the point of view of common sense the fluxes in $\mathbf{\Pi}_2, \dots, \mathbf{\Pi}_M$ depend on physical phenomena which take place on the ends of these pipes. For example, it can happen that some pipes have either plugs or pumps (i.e. the flux is either equal to zero, or is prescribed), others are open and the liquid can run–off (i.e. the pressure at infinity must be equal to the atmospheric one). There are also physical phenomena which lead to non–linear asymptotic conditions at infinity. For example, the pipes are supplied either by a semi–permeable (or porous) membrane or by a one direction check valve at a very long distance. Thus, the liquid can flow through such pipe in one direction only and only if the pressure in it is larger than certain given constant pressure p_0 and the asymptotic conditions of variational inequality type ought to be posed at infinity. It is self–understood, that it is not possible to list all sensible physical situations.

Stokes and Navier–Stokes equations supplied by asymptotic conditions at infinity in general setting are studied in forthcoming papers by S.A. Nazarov and the author [38], [39]. In [38], [39] we consider both linear and nonlinear conditions at infinity (for Stokes and Navier-Stokes systems) and the last includes also the conditions of variational inequality type. Large attention in these papers is given to examples where the asymptotic conditions at infinity are taken based on physical reasonings and are included in the proposed scheme. We consider the linear asymptotic conditions which correspond to pipes either with plugs or pumps, open drains or long tubes connected by a taper pipe. We also study the asymptotic conditions at infinity, connecting the coefficients in the asymptotics by nonlinear relations, in particular, with the help of algebraic variational inequalities. In particular, we consider a check valve, which permits the liquid to flow in a pipe only in one direction.

Let us describe the main points of the reasonings for the linear Stokes equations. The Stokes operator is considered on a weighted function space with the detached asymptotics; in other words, a solution $(u,\ p)$ is represented as a sum of Poiseuille flows and constant pressures in each outlet while a remainder in such representation is characterized by an exponential decay at infinity. Based on the theory of general elliptic problems in cylindric domains (see [20], [27], [40], etc.), it is shown that the Stokes operator with such domain is a Fredholm epimorphism with a M–dimensional kernel (M is the number of the outlets to infinity). In order to remove, if it is possible, the appeared non–uniqueness of the solution, the Stokes equations are supplied with additional conditions at infinity, i.e. they are associated with a certain system of M linear algebraic equations, relating the constants in the asymptotic representation of a solution. The choice of this system is, in general, arbitrary. However, there exists an approach imitating the posing of boundary conditions in domains with smooth compact boundaries and based on a suitable Green's formula. The point is that the usual for the Stokes problem Green's formula ceases to be true for vector fields from the mentioned function space with the detached asymptotics – there appear additional terms: the quadratic form of coefficients in the asymptotic representation of the solution. Such formula we name " generalized Green's formula "(cf. [40]). The needed algebraic system is extracted from this generalized Green's formula just in the same way as stable, natural and mixed boundary conditions are obtained from the usual Green's formula (the terminology is taken from [26]). In addition, if the Green's formula is symmetric, then the Stokes problem with the corresponding boundary conditions turns out to be formally self–adjoint.

In order to form an integral characteristic of the Stokes problem in the domain Ω with outlets to infinity, namely, the unitary $M \times M$–scattering matrix $\mathcal{S}_\Omega$, we use the standard constructions of the scattering theory . With the help of the transform, which is similar to the Kelly's transform, from $\mathcal{S}_\Omega$ it is possible to obtain another integral characteristic – a symmetric $M \times M$–matrix $\mathcal{Q}_\Omega$. This matrix, named the pressure distribution matrix, has the clear physical interpretation and plays the primary role in the investigation of problems with the asymptotic conditions at infinity: the conditions of unique solvability of the problem and dimensions of kernel and cokernel of its operator are described in terms of this matrix.

The outlined analogy with the usual boundary value problems extends further: in the self–adjoint case it is possible to construct the functionals for which the problems with the asymptotic conditions at infinity are equivalent to the minimization problems in a class of functions with the detachted asymptotics. We mention in this connection that the possibility of variational formulation is very important, in particular, for numerical purposes (e.g. [12]).

The questions of approximation of semi-infinite cylinders by long finite ones and the asymptotically exact estimates for the corresponding solutions are studied in forthcoming papers of M. Specovius-Neugebauer.

5.3. Navier-Stokes equations combined with other phenomena

Having in mind practical applications, it is important to study the equations of fluid motion assuming either additional physical phenomena, or more complicated than Navier-Stokes equations of fluid motion. We have no possibility to discuss here in detail such problems and we only refer to

(a) [45], [46], [47], [57], [34], [1], [14] for the results concerning the Navier-Stokes equations in domains with noncompact free boundaries governed by surface tension (so-called coating flows);

(b) [41] for the results concerning a flow of compressible fluid filling a cylinder, subject to compact support body forces and to a prescribed flux of the momentum;

(c) [51] for the results concerning the motion of second-grade and Oldroyd-B fluids in domains with cylindric outlets to infinity.

References

[1] Abergel, F., Bona, J.L., A mathematical theory for viscous free–surface flows over a perturbed plane, Arch. Rational Mech. Anal., **118**, 71–93 (1992).

[2] Agmon, S., Douglis, A., Nirenberg, L., Estimates near the boundary for solutions of elliptic partial differential equations satisfying general boundary conditions II, Commun. Pure Appl. Math., **17**, 35–92 (1964).

[3] Amick, C.J., Steady solutions of the Navier–Stokes equations in unbounded channels and pipes, Ann. Scuola Norm. Sup. Pisa, **4**, 473–513 (1977).

[4] Amick, C.J., Properties of steady Navier–Stokes solutions for certain unbounded channels and pipes, Nonlinear analysis, Theory, Meth., Appl., **2**, 689–720 (1978).

[5] Babenko, K.I., On stationary solutions of the problem of flow past a body of a viscous incompressible fluid, Math. USSR Sbornik, **20** , No 1, 1–25 (1973).

[6] Bogovskii, M.E., Solutions of some problems of vector analysis related to operators *div* and *grad*, Proc. Semin. S.L. Sobolev, **1**, 5–40 (1980) (in Russian).

[7] Borchers, W., Pileckas, K., Existence, uniqueness and asymptotics of steady jets, Arch. Rational Mech. Anal., **120**, 1–49 (1992).

[8] Borchers, W., Galdi, G.P., Pileckas, K., On the uniqueness of Leray-Hopf solutions for the flow through an aperture, Arch. Rational Mech. Anal., **122**, 19–33 (1993).

[9] Galdi, G.P., An introduction to the mathematical theory of the Navier–Stokes equations, Springer Tracts in Natural Philosophy **38, 39** (1994).

[10] Galdi, G.P., Padula, M., Solonnikov, V.A., Existence, uniqueness and asymptotic behavior of solutions of steady-state Navier-Stokes equations in a plane aperture domain (to appear).

[11] Heywood, J.G., On uniqueness questions in the theory of viscous flow, Acta. Math., **136**, 61–102 (1976).

[12] Heywood, J.G., Rannacher, R., Turek, S., Artificial boundaries and flux and pressure drop conditions for the incompressible Navier–Stokes equations, Preprint Nr. 681, Universität Heidelberg (1992).

[13] Finn, R., On the exterior stationary problem for the Navier–Stokes equations and associated perturbation problems, Arch. Rational Mech. Anal. **19**, 363–406 (1965).

[14] Friedman, A., Velázquez, J.J.L., The analysis of coating flows in a strip, J. Diff. Equations, **121**, No.1, 134–182 (1995).

[15] Horgan, C.O., Plane steady flows and energy estimates for the Navier-Stokes equations, Arch. Rational Mech. Anal., **68**, 359–381 (1978).

[16] Horgan, C.O., Wheeler, L.T., Spatial decay estimates for the Navier-Stokes equations with application to the problem of entry flow, SIAM J. Appl. Math., **35**, 97–116 (1978).

[17] Kapitanskii, L.V., Coincidence of the spaces $\overset{o}{J}{}_2^1(\Omega)$ and $\hat{J}_2^1(\Omega)$ for plane domains Ω having exits at infinity, Zapiski Nauch. Semin. LOMI, **110**, 74–81 (1981). English Transl.: J. Sov. Math., **25**, 850–855 (1984).

[18] Kapitanskii, L.V., Stationary solutions of the Navier–Stokes equations in periodic tubes, Zapiski Nauch. Semin. LOMI, **115**, 104–113 (1982). English Transl.: J. Sov. Math., **28**, 689–695 (1983).

[19] Kapitanskii, L.V., Pileckas, K., On spaces of solenoidal vector fields and boundary value problems for the Navier-Stokes equations in domains with noncompact boundaries,Trudy Mat. Inst. Steklov, **159**, 5–36 (1983). English Transl.: Proc. Math. Inst. Steklov, **159**, issue 2, 3–34 (1984).

[20] Kondrat'ev, V.A., Boundary value problems for elliptic equations in domains with conical or corner points, Trudy Moskov. Mat Obshch., **16**, 209–292 (1967). English Transl.: in Trans. Moscow Math. Soc., **16** (1967).

[21] Ladyzhenskaya, O.A., The mathematical theory of viscous incompressible flow, Gordon and Breach, New York, London, Paris (1969).

[22] Ladyzhenskaya, O.A., Solonnnikov, V.A., On some problems of vector analysis and generalized formulations of boundary value problems for the Navier–Stokes equations, Zapiski Nauchn. Sem. LOMI, **59**, 81–116 (1976). English Transl.: J. Sov. Math., **10**, No. 2, 257–285 (1978).

[23] Ladyzhenskaya, O.A., Solonnnikov, V.A., On the solvability of boundary value problems for the Navier-Stokes equations in regions with noncompact boundaries, Vestnik Leningrad. Univ., **13** (Ser. Mat. Mekh. Astr. vyp. 3), 39–47 (1977). English Transl.: Vestnik Leningrad Univ. Math., **10**, 271–280 (1982).

[24] Ladyzhenskaya, O.A., Solonnnikov, V.A., Determination of the solutions of boundary value problems for stationary Stokes and Navier-Stokes equations having an unbounded Dirichlet integral, Zapiski Nauchn. Sem. LOMI, **96**, 117–160 (1980). English Transl.: J. Sov. Math., **21**, No.5, 728–761 (1983).

[25] Ladyzhenskaya, O.A., Solonnnikov, V.A., On initial–boundary value problems for the linearized Navier–Stokes system in domains with noncompact boundaries., Trudy Mat. Inst. Steklov, **159**, (1983). English Transl.: Proc. Math. Inst. Steklov, **159**, issue 2, 35–40 (1984).

[26] Lions, J.L., Magenes, E., Nonhomogeneous boundary value problems I, Springer Verlag, Berlin (1972).

[27] Maz'ya, V.G., Plamenevskii, B.A., Estimates in L_p and Hölder classes and the Miranda-Agmon maximum principle for solutions of elliptic boundary value problems in domains with singular points on the boundary, Math. Nachr., **81**, 25–82 (1978). English Transl.: Amer. Math. Soc. Transl., **123** (2), 1–56 (1984).

[28] Maz'ya, V.G., Plamenevskii, B.A., The coefficients in the asymptotic expansion of the solutions of elliptic boundary value problems in a cone, Zapiski Nauchn. Sem. LOMI, **52**, 110–127 (1975). English Transl.: J. Sov. Math., **9**, No. 5, 750–764 (1978).

[29] Maz'ya, V.G., Plamenevskii, B.A., On the coefficients in the asymptotics of solutions of elliptic boundary value problems in domains with conical points, Math. Nachr., **76**, 29–60 (1977). English Transl.: Amer. Math. Soc. Transl., **123** (2), 57–88 (1984).

[30] Nazarov, S.A., The structure of the solutions of elliptic boundary value problems in thin domains, Vestnik Leningrad. Univ., **7** (Ser. Mat. Mekh. Astr., vyp. 2), 65–68 (1982). English Transl.: Vestnik Leningrad. Univ. Math., **15** (1983).

[31] Nazarov, S.A., On the two–dimensional aperture problem for Navier–Stokes equations (to appear).

[32] Nazarov, S.A., Asymptotic solution of the Navier–Stokes problem on the flow of a thin layer of fluid, Sibirsk. Mat. Zh., **31**, No. 2, 131–144 (1990). English Transl.: Siberian Math. Journal, **31**, 296-307 (1990).

[33] Nazarov, S.A., Pileckas, K., On the behaviour of solutions of the Stokes and Navier–Stokes systems in domains with periodically varying section, Trudy Mat. Inst. Steklov, **159**, 137–149 (1983). English Transl.: Proc. Math. Inst. Steklov, **159**, issue 2, 141–154 (1984).

[34] Nazarov, S.A., Pileckas, K., On noncompact free boundary problems for the plane stationary Navier–Stokes equations, J. für reine und angewandte Math., **438**, 103–141 (1993).

[35] Nazarov, S.A., Pileckas, K., The Reynolds flow of a fluid in a thin three–dimesional channel, Litovskii Mat. Sb., **30**, 772–783 (1990). English Transl.: Lithuanian Math. J., **30** (1990).

[36] Nazarov, S.A., Pileckas, K., Asymptotics of solutions to Stokes and Navier–Stokes equations in domains with paraboloidal outlets to infinity, Rend. Sem. Math. Univ. Padova (to appear).

[37] Nazarov, S.A., Pileckas, K., Fredholm property of the Neumann problem in domains with layer-like outlet to infinity, Algebra i Analis (1996). English Transl.: St.-Petersburg Math. J. (1996) (to appear).

[38] Nazarov, S.A., Pileckas, K., Asymptotic conditions at infinity for the Stokes and Navier–Stokes problems in domains with cylindric outlets to infinity. I. Linear problems (to appear).

[39] Nazarov, S.A., Pileckas, K., Asymptotic conditions at infinity for the Stokes and Navier–Stokes problems in domains with cylindric outlets to infinity. II. Nonlinear problems (to appear).

[40] Nazarov, S.A., Plamenevskii, B.A., Elliptic boundary value problems in domains with piecewise smooth boundary, Valter de Gruyter and Co., Berlin (1993).

[41] Padula, M., Pileckas, K., On the existence and asymptotical behaviour of a steady flow of a viscous barotropic gas in a pipe, Annali di Mathem. Pura ed Appl. (to appear).

[42] Pileckas, K., Three–dimensional solenoidal vectors, Zapiski Nauchn. Sem. LOMI, **96**, 237–239 (1980). English Transl.: J. Sov. Math., **21**, 821–823 (1983).

[43] Pileckas, K., Existence of solutions for the Navier–Stokes equations having an infinite dissipation of energy, in a class of domains with noncompact boundaries, Zapiski Nauchn. Sem. LOMI, **110**, 180–202 (1981). English Transl.: J. Sov. Math., **25**, No.1, 932–947 (1984).

[44] Pileckas, K., On spaces of solenoidal vectors, Trudy Mat. Inst. Steklov, **159**, 137–149 (1983). English Transl.: Proc. Math. Inst. Steklov, **159**, issue 2, 141–154 (1984).

[45] Pileckas, K., On the solvability of certain problem of a plane motion of viscous incompressible liquid with a noncompact free boundary, Zapiski Nauchn. Sem. LOMI, **110**, 174–179 (1981). English Transl.: J. Sov. Math., **25**, No.1, 927–931 (1984).

[46] Pileckas, K., On the problem of motion of heavy viscous incompressible fluid with noncompact free boundary, Litovskii Mat. Sb., **28**, 315-333 (1988). English Transl.: Lithuanian Math. J., **28**, No. 2 (1988).

[47] Pileckas, K., On plane motion of a viscous incompressible cappilary liquid with a noncompact free boundary, Arch. Mech., **41**, No. 2-3, 329-342 (1989).

[48] Pileckas, K., Weighted L^q–solvability for the steady Stokes system in domains with noncompact boundaries, M^3AS: Math. Models and Methods in Applied Sciences, **6**, No. 1, 97-136 (1996).

[49] Pileckas, K., Classical solvability and uniform estimates for the steady Stokes system in domains with noncompact boundaries, M^3AS: Math. Models and Methods in Applied Sciences, **6**, No. 2, (1996).

[50] Pileckas, K., Strong solutions of the steady nonlinear Navier–Stokes system in domains with exits to infinity, Rendiconti del Seminario Matematico dell' Universita di Padova (to appear).

[51] Pileckas, K., Sequeira, A., Videman, J.H., A note on steady flows of non-newtonian fluids in channels and pipes (to appear).

[52] Solonnikov, V.A., On the boundary value problems for systems elliptic in the sense of A. Douglis, L. Nirenberg, I. Izv. Akad. Nauk SSSR, Ser. Mat., **28**, 665–706 (1964); II. Trudy Mat. Inst. Steklov, **92**, 233–297 (1966). English Transl.: I. Amer. Math. Soc. Transl., **56** (2), 192–232 (1966), II. Proc. Steklov Inst. Math., **92**, 269–333 (1966).

[53] Solonnikov, V.A., On the solvability of boundary and initial–boundary value problems for the Navier–Stokes system in domains with noncompact boundaries, Pacific J. Math., **93**, No.2, 443–458 (1981).

[54] Solonnikov, V.A., On solutions of stationary Navier–Stokes equations with an infinite Dirichlet integral, Zapiski Nauchn. Sem. LOMI, **115**, 257–263 (1982). English Transl.: J. Sov. Math., **28**, No.5, 792–799 (1985).

[55] Solonnikov, V.A., Stokes and Navier–Stokes equations in domains with noncompact boundaries, College de France Seminar, **4**, 240-349 (1983).

[56] Solonnikov, V.A., Boundary and initial–boundary value problems for the Navier–Stokes equations in domains with noncompact boundaries, Math. Topics in Fluid Mechanics, Pitman Research Notes in Mathematics Series, **274** (eds. J.F. Rodrigues, A. Sequeira), 117–162 (1991).

[57] Solonnikov, V.A., Solvability of the problem of effluence of a viscous incompressible fluid into an open basin, Trudy Mat. Inst. Steklov, **179**, 193–225 (1989). English Transl.: Proc. Math. Inst. Steklov, **179**, issue 2, 193–225 (1989).

[58] Solonnikov, V.A., Pileckas, K., Certain spaces of solenoidal vectors and the solvability of the boundary value problem for the Navier-Stokes system of equations in domains with noncompact boundaries, Zapiski Nauchn. Sem. LOMI, **73**, 136–151 (1977). English Transl.: J. Sov. Math., **34**, No.6 , 2101–2111 (1986).

[59] Stein, E.M., Singular integrals and differentiability properties of functions, Princeton University Press (1970).

[60] Toupin, R.A., Saint–Venant's principle, Arch. Rational Mech. Anal., **18**, 83–96 (1965).

Konstantin Pileckas
University of Paderborn
FB 17
D-33095 Paderborn
Germany

K.R. RAJAGOPAL

An Introduction to Mixture Theory

1. Introduction

Here, I shall try in the course of four lectures to present the participants of the winter school the salient aspects of mixture theory. Constraint of space and time do not allow the luxury of a presentation that is self-contained and complete, and as the subject matter of the lectures spans the whole gamut of gaseous mixtures, bubbly liquids, liquid mixtures, particulate suspensions, solids infused with fluids, amongst other mixtures, it should be self evident that the subject matter cannot be confined to a mere four lectures. Moreover, as most of the participants of the winter school are not experts in mixture theory, the task of presenting the material at any depth becomes all the more onerous. I refer the interested participant to the recent book by Rajagopal and Tao [1], with the caveat that the book is incomplete and not self-contained and is not meant as a text, requiring of the reader the arduous task of referring to the primary sources. The various appendices in Rational Thermodynamics [2] and the review articles by Atkin and Craine [3], Bowen [4], Bedford and Drumheller [5], Passman, Nunziato and Walsh [6] can be used to fill in the gaps, nay chasms in these lectures; however they themselves cover but few aspects of mixture theory and moreover much progress has been made in the theory since the above papers appeared.

There have been various approaches that have been advanced for modelling mixtures. Here, I shall content myself with a continuum approach to modelling mixtures which presumes that the various constituents of the mixture are sufficiently well distributed throughout the configuration occupied by the constituents to render such an approach reasonable within the context of homogenization. The earliest phenomenological theory of mixture was proposed by Truesdell [7], [8]. Truesdell was influenced by the seminal works of Maxwell [9] and Chapman and Cowling [10] in the kinetic theory of gases and the early work on diffusion by Fick [11]. Early papers from a statistical point of view that deal with mixtures are those due to Kirkwood and Buff [12], Pearson and Rushbrooke [13], Mazo [14], and Bearman and Kirkwood [15]. Copious references to statistical approach to mixtures can be found in Mansoori and Matteoli [16]. Another early source for a continuum theory of mixtures in the authoritative tome by Truesdell and Toupin [17].

Before we get into a discussion of continuum mixture theory, it would be appropriate to discuss how mixtures were studied before its advent by the use of Fick's or Darcy's equation or some adhoc generalization of it. While the linear relationship between the mass flux and the concentration gradient has been exalted to status of a "law", it is at best an appropriate approximation of the balance of linear momentum (see Bowen [4]), providing no insight into the state of stress or strain in the two constituents. The same can be said of "Darcy's law" that relates the flux, of a constituent diffusing through another, to the pressure gradient. Moreover, both these relations ignore the inertia of the constituents, and these "laws" (correlations) capture the diffusion of one constituent through another only under very special conditions. There have been numerous efforts to generalize these "laws" to extend their range of validity, but all these attempts are generally quite ad hoc. It is not my aim here to impugn "Darcy's law" or "Fick's law" (Fick [11], Darcy [18]), to the contrary the purpose of the above remarks is to emphasize that these "laws" serve their purpose admirably provided the user does not employ them beyond the realms of their validity, a fact that the early proponents of the use of these "laws" were well cognizant of. The final section of these notes is devoted to a discussion of the status of Darcy's law within the context of the theory of mixtures.

2. Kinematics and Basic Balance Laws

The basic premise of the continuum theory of mixtures is that each point in the space occupied by the mixture is occupied by a particle belonging to each constituent. Such a co-occupancy by the various constituents that comprise the mixture can only be given meaning within the context of an appropriate homogenization, and is in no way different from the spirit in which a single constituent body is approximated as a continuum for such a body is after all full of voids at the microscopic level, the body being homogenized in a manner that covers these voids. Thus, in order for a continuum theory of mixture to be applicable, the various constituents should be distributed in such a manner that each of them could in their own right be considered as a single continuum. This point cannot be over emphasized for otherwise the theory would be put to use beyond its range of applicability. Henceforth, we shall assume that the mixture is such that the basic assumption of the continuum theory of mixtures is valid.

In general, as the mixture deforms, there is relative motion between the constituents. Let us denote by $\boldsymbol{X}^i$, $i = 1, \dots, n$, a typical material point belonging to the reference configuration of the ith constituent. Let $\boldsymbol{x}$ denote a typical point in the configuration at time t occupied by the mixture. Then, we can define one-to-one mappings $\boldsymbol{\chi}^i$ that defines the motion of the ith constituent through

$$\boldsymbol{x} = \boldsymbol{\chi}^i(\boldsymbol{X}^i, t) \,. \tag{2-1}$$

Here, we shall not follow the usual Einstein summation convention and summation is not intended on repeated indices unless it is explicitly stated. We shall assume that the mappings $\boldsymbol{\chi}^i$s are sufficiently smooth to render all the differential operations meaningful.

The velocity of particles belonging to each constituent is defined through

$$\boldsymbol{v}^i = \frac{\partial \boldsymbol{\chi}^i}{\partial t}, \quad i = 1, \ldots, n. \tag{2-2}$$

Given the velocity of the particles of each constituent, we can define a mean velocity for the mixture in a variety of ways. Here, we shall define a mass averaged mean velocity for the mixture. Let ϱ_R^i denote the density of the ith constituent in its reference configuration, and let ϱ^i denote the density of the constituent in the current configuration (i.e., configuration at time t). We define the total density of the mixture ϱ through

$$\varrho = \sum_{i=1}^{n} \varrho^i, \tag{2-3}$$

that is completely in keeping with our basic hypothesis of co-existence. We define the mean velocity $\boldsymbol{v}$ of the mixture through

$$\boldsymbol{v} = \frac{1}{\varrho} \sum_{i=1}^{n} \varrho^i \boldsymbol{v}^i. \tag{2-4}$$

The difference between the mean velocity $\boldsymbol{v}$ and the constituent velocity $\boldsymbol{v}^i$ provides a measure of how the ith constituent diffuses with respect to the mixture as a whole and is thus referred to as the "diffusion velocity" and is defined through

$$\boldsymbol{u}^i := \boldsymbol{v}^i - \boldsymbol{v}. \tag{2-5}$$

In addition to the usual lagrangian time derivative that is based on following particles belonging to each constituent, it is also possible to define time derivatives of physical quantities with respect to an observer who is moving with the mean velocity of the mixture. We use the notation (see Bowen [4])

$$\grave{f}^i := \frac{D^{(i)} f}{Dt} := \frac{\partial f}{\partial t} \left(\boldsymbol{\chi}^i \left(\boldsymbol{X}^i, t \right), t \right) \Big|_{\boldsymbol{X}^i \text{held fixed}} = \frac{\partial f}{\partial t} + \left[\frac{\partial f}{\partial \boldsymbol{x}} \right] \boldsymbol{v}^i, \tag{2-6}$$

where f is a scalar, vector or tensor, with the operation $[\frac{\partial f}{\partial \boldsymbol{x}}]\boldsymbol{v}^i$ taking on the appropriate sense for each of these cases. We also define the time derivative with respect to an observer moving with the mean velocity of the mixture through

$$\dot{f} := \frac{Df}{Dt} = \frac{\partial f}{\partial t} + \left[\frac{\partial f}{\partial \boldsymbol{x}} \right] \boldsymbol{v}. \tag{2-7}$$

The deformation gradient $\boldsymbol{F}^i$ associated with the ith constituent is defined through

$$\boldsymbol{F}^i := \frac{\partial \boldsymbol{\chi}^i}{\partial \boldsymbol{X}^i}, \quad i = 1, \ldots, n. \tag{2-8}$$

We shall assume that

$$\det \boldsymbol{F}^i \neq 0\,, \tag{2-9}$$

and without loss of generality, we shall assume that $\det \boldsymbol{F}^i > 0$.

The acceleration $\boldsymbol{a}^i$ of the ith constituent is defined through

$$\boldsymbol{a}^i = \frac{\partial^2 \boldsymbol{\chi}^i}{\partial t^2}\,, \tag{2-10}$$

while the velocity gradient $\boldsymbol{L}^i$ of the ith constituent is defined through

$$\boldsymbol{L}^i = \frac{\partial \boldsymbol{v}^i}{\partial \boldsymbol{x}}\,. \tag{2-11}$$

It is then a simple matter to show that

$$\boldsymbol{L}^i = \dot{\boldsymbol{F}}^i \left(\boldsymbol{F}^i\right)^{-1}. \tag{2-12}$$

The left and right Cauchy-Green Stretch tensors are defined respectively through

$$\boldsymbol{B}^i = \boldsymbol{F}^i \left(\boldsymbol{F}^i\right)^{\top}, \quad i = 1, \ldots, n\,, \tag{2-13}$$

$$\boldsymbol{C}^i = \left(\boldsymbol{F}^i\right)^{\top} \boldsymbol{F}^i, \quad i = 1, \ldots, n\,, \tag{2-14}$$

and the Green-St. Venant and Almansi-Hamel strain tensors are defined through

$$\boldsymbol{E}^i = \frac{1}{2}\left(\boldsymbol{C}^i - \boldsymbol{1}\right), \quad i = 1, \ldots, n\,, \tag{2-15}$$

$$\boldsymbol{e}^i = \frac{1}{2}\left(\boldsymbol{1} - (\boldsymbol{B}^i)^{-1}\right), \quad i = 1, \ldots, n\,. \tag{2-16}$$

The symmetric and skew parts of the velocity gradient are defined through

$$\boldsymbol{D}^i = \frac{1}{2}\left(\boldsymbol{L}^i + (\boldsymbol{L}^i)^{\top}\right), \tag{2-17}$$

$$\boldsymbol{W}^i = \frac{1}{2}\left(\boldsymbol{L}^i - (\boldsymbol{L}^i)^{\top}\right), \tag{2-18}$$

and the vorticity $\boldsymbol{\omega}^i$ is the axial vector associated with the spin tensor $\boldsymbol{W}^i$.

For any tensor $\boldsymbol{A}$, the principal invariants are given by

$$I_1 = \operatorname{tr} \boldsymbol{A}\,, \quad I_2 = \frac{1}{2}\left[(\operatorname{tr} \boldsymbol{A})^2 - \operatorname{tr} \boldsymbol{A}^2\right], \quad I_3 = \det \boldsymbol{A}\,. \tag{2-19}$$

Kinematical quantities based on a mass averaging for the mixture as a whole, similar to the mixture velocity, can be defined in a straightforward manner. We shall not do so here, but a detailed treatment can be found in Bowen [4] and Atkin and Craine [3].

As is customary in the development of the mechanics of a single continua, we shall introduce the notion of surface tractions. Let s denote the surface containing a point $\boldsymbol{x}$ belonging to the mixture, and let $\boldsymbol{n}_s$ denote the unit outward normal to the surface. Let $\boldsymbol{t}^i$ denote the partial traction vector associated with the ith constituent. Proceeding in a manner identical to that for a single continua, we can show the existence of a stress tensor $\boldsymbol{T}^i$ (see Truesdell [19]) such that

$$\boldsymbol{t}^i = \left(\boldsymbol{T}^i\right)^{\top} \boldsymbol{n} . \tag{2-20}$$

We shall define the total traction $\boldsymbol{t}$ and the total stress tensor $\boldsymbol{T}$ for the mixture through

$$\boldsymbol{t} = \sum_{i=1}^{n} \boldsymbol{t}^i , \quad \boldsymbol{T} = \sum_{i=1}^{n} \boldsymbol{T}^i . \tag{2-21}$$

It is appropriate to caution the reader that the representation for the total stress associated with the mixture has been the subject of some controversy (see Bowen [4], Green and Naghdi [20], Truesdell [2]). It might also be fitting to mention that, for mixtures of gases and of liquids, expressions for the partial tractions, and the total traction, have been derived within the purview of statistical methods.

Balance of Mass

Mixture theory can take into account the chemical reactions that take place between the constituents which can lead to interconversion amongst the constituents (see Samohyl [21]). We shall denote by m^i the mass supply to the ith constituent due to the reactions between the other constituents. Then, for any part $\mathcal{P}$ of the body, the balance of mass for the ith constituent takes the form

$$\frac{\partial}{\partial t} \int_{\mathcal{P}} \varrho^i dv + \int_{\partial \mathcal{P}} \varrho^i \boldsymbol{v}^i \cdot \boldsymbol{da} = \int_{\mathcal{P}} m^i dv , \tag{2-22}$$

where $\partial \mathcal{P}$ denotes the boundary of the part $\mathcal{P}$. Then by standard arguments, if the integrands are continuous, we obtain the local form of the balance of mass

$$\grave{\varrho}^i + \varrho^i \operatorname{div} \boldsymbol{v}^i = m^i . \tag{2-23}$$

In addition to the balance of mass for the individual constituents, we can also write down the balance of mass for the mixture as a whole, namely

$$\frac{\partial}{\partial t} \int_{\mathcal{P}} \varrho dv + \int_{\partial \mathcal{P}} \varrho \boldsymbol{v} \cdot \boldsymbol{da} = 0 , \tag{2-24}$$

which gives the local form

$$\dot{\varrho} + \varrho \operatorname{div} \boldsymbol{v} = 0\,, \tag{2-25}$$

which has the same structure as that for a single constituent.

On summing (2-23) over i, and comparing with (2-25) leads to

$$\sum_{i=1}^{n} m^i = 0\,, \tag{2-26}$$

which affirms that there is no net production of mass.

The above formulations of the balance of mass are from the eulerian viewpoint. We shall also record the lagrangian form of the balance of mass as we shall find it convenient to express the balance of mass for solid constituents in that manner:

$$\frac{d}{dt}\int_{\Omega} \varrho^i dV = \int_{\Omega} m^i dV\,, \tag{2-27}$$

which leads to the local form

$$\dot{\overline{\varrho^i \det \boldsymbol{F}^i}} = m^i \det \boldsymbol{F}^i\,. \tag{2-28}$$

Balance of Linear Momentum

The integral form for the balance of linear momentum for the ith constituent takes the form

$$\frac{\partial}{\partial t}\int_{\mathcal{P}} \varrho^i \boldsymbol{v}^i dv + \int_{\partial\mathcal{P}} \varrho^i \boldsymbol{v}^i (\boldsymbol{v}^i \cdot \boldsymbol{da}) = \int_{\partial\mathcal{P}} \boldsymbol{\sigma}^{\top} \boldsymbol{da} + \int_{\mathcal{P}} \left(\varrho^i \boldsymbol{b}^i + \boldsymbol{m}^i + m^i \boldsymbol{v}^i\right) dv\,, \tag{2-29}$$

where $\boldsymbol{b}^i$ is the specific external body force and $\boldsymbol{m}^i$ is the momentum supply to the ith constituent due to its interaction with the other constituents, and $m^i \boldsymbol{v}^i$ is the momentum associated with the mass m^i of the ith constituent that has been created. By standard arguments based on the smoothness of the integrands we can establish the differential form of (2-29) as

$$\frac{\partial}{\partial t}(\varrho^i \boldsymbol{v}^i) + \operatorname{div}(\varrho^i \boldsymbol{v}^i \otimes \boldsymbol{v}^i) = \operatorname{div}(\boldsymbol{\sigma}^i)^{\top} + \varrho^i \boldsymbol{b}^i + \boldsymbol{m}^i + m^i \boldsymbol{v}^i\,. \tag{2-30}$$

It can be shown that (2-30) can be expressed as

$$\varrho^i \frac{D^{(i)} \boldsymbol{v}^i}{Dt} = \operatorname{div}(\boldsymbol{\sigma}^i)^{\top} + \varrho^i \boldsymbol{b}^i + \boldsymbol{m}^i\,. \tag{2-31}$$

On summing (2-31) over i, we obtain

$$\varrho\dot{\boldsymbol{v}} = \operatorname{div}\left[\boldsymbol{\sigma}^{\top} - \sum_{i=1}^{n}(\varrho^i \boldsymbol{u}^i \otimes \boldsymbol{u}^i)\right] + \sum_{i=1}^{n}(\boldsymbol{m}^i + m^i \boldsymbol{v}^i) + \varrho \boldsymbol{b}, \tag{2-32}$$

where

$$\boldsymbol{b} = \frac{1}{\varrho}\sum_{i=1}^{n} \varrho^i \boldsymbol{b}^i . \tag{2-33}$$

We see that (2-32) can be put in the usual form for the balance of linear momentum for a single constituent, if we set

$$\hat{\boldsymbol{\sigma}} = \boldsymbol{\sigma} - \sum_{i=1}^{n} \varrho^i \boldsymbol{u}^i \otimes \boldsymbol{u}^i , \tag{2-34}$$

provided

$$\sum(\boldsymbol{m}^i + m^i \boldsymbol{v}^i) = 0 . \tag{2-35}$$

The above fact has prompted some to define (see Truesdell [7]) the total stress tensor through (2-34) rather than (2-21). Statistical arguments have been advanced for determining the relationship between the partial stresses and the total stress of the mixture but we shall not get into them here. In the rest of these lectures, we shall define the total stress through (2-21).

A few remarks about the interactive body force $\boldsymbol{m}^i$ are warranted as they play a crucial role in the theory. As one constituent diffuses through the others, various interactions come into play, a detailed discussion of the same can be found in Johnson, Massoudi and Rajagopal [22]. The most important of these are drag, lift, virtual mass effect, history effects (Basset forces), and the effect of relative spinning (Magnus effect). Depending on the problem under consideration one, some or all of these effects would have to be taken into account.

Balance of Angular Momentum

The balance of moment of linear momentum (angular momentum) of the ith constituent takes the form

$$\begin{aligned} &\frac{\partial}{\partial t}\int_{\mathcal{P}} \boldsymbol{x} \times \varrho^i \boldsymbol{v}^i dv + \int_{\partial\mathcal{P}} \boldsymbol{x} \times \varrho^i \boldsymbol{v}^i (\boldsymbol{v}^i \cdot \boldsymbol{da}) = \\ &= \int_{\partial\mathcal{P}} \boldsymbol{x} \times \sigma^{\top} \boldsymbol{da} + \int_{\mathcal{P}} \left[\boldsymbol{x} \times (\varrho^i \boldsymbol{b}^i + \boldsymbol{m}^i + m^i \boldsymbol{v}^i) + \boldsymbol{M}^i\right] dv , \end{aligned} \tag{2-36}$$

where $\boldsymbol{M}^i$ is the angular momentum supply to the ith constituent.

In the case of a single constituent, if there are no body couples, the balance of angular momentum leads to the conclusion that the Cauchy stress tensor is symmetric. However, in the case of mixtures, we find

$$\boldsymbol{\varepsilon M}^i = (\boldsymbol{\sigma}^i)^\top - \boldsymbol{\sigma}^i , \tag{2-37}$$

where $\boldsymbol{\varepsilon}$ is the alternator. When there is no angular momentum supply to the ith constituent, we obtain the familiar result that the partial stress associated with the ith component is symmetric.

If

$$\sum_{i=1}^{n} \boldsymbol{\varepsilon M}^i = 0 , \tag{2-38}$$

then it follows from (2-36), on summing over i and using (2-38) and smoothness arguments, that

$$\overline{\boldsymbol{x} \times \varrho \boldsymbol{v}}^{\,\cdot} = \operatorname{div}(\boldsymbol{x} \times \boldsymbol{\sigma}^\top) + \boldsymbol{x} \times \varrho \boldsymbol{b} . \tag{2-39}$$

Conservation of Energy

Let ε^i, $\boldsymbol{q}^i$, r^i, ε^i_s denote the specific internal energy, the heat flux vector, the radiant heating and the energy production, respectively, of the ith constituent. The conservation of energy for the ith constituent then takes the form

$$\begin{aligned} &\frac{\partial}{\partial t} \int_{\mathcal{P}} \varrho^i (\varepsilon^i + \frac{1}{2} \boldsymbol{v}^i \cdot \boldsymbol{v}^i) dv + \int_{\mathcal{P}} \varrho^i (\varepsilon^i + \frac{1}{2} \boldsymbol{v}^i \cdot \boldsymbol{v}^i) \boldsymbol{v}^i dv = \\ &= \int_{\partial \mathcal{P}} (\boldsymbol{\sigma}^i \boldsymbol{v}^i - \boldsymbol{q}^i) \cdot d\boldsymbol{a} + \int_{\mathcal{P}} (\varrho^i r^i + \varrho^i \boldsymbol{v}^i \cdot \boldsymbol{b}^i) dv + \\ &+ \int_{\mathcal{P}} \left[\varepsilon^i_s + m^i (\varepsilon^i + \frac{1}{2} \boldsymbol{v}^i \cdot \boldsymbol{v}^i) + \boldsymbol{m}^i \cdot \boldsymbol{v}^i \right] dv . \end{aligned} \tag{2-40}$$

The last integral on the right hand side denotes the energy that is supplied to the part $\mathcal{P}$, the term $m^i(\varepsilon^i + \frac{1}{2}\boldsymbol{v}^i \cdot \boldsymbol{v}^i)$ denoting the part due to the mass generation of the ith constituent and the term $\boldsymbol{m}^i \cdot \boldsymbol{v}^i$ the contribution due the momentum production of the ith constituent. Using standard arguments we obtain the local form of the above equation

$$\begin{aligned} &\frac{\partial}{\partial t} \left[\varrho^i (\varepsilon^i + \frac{1}{2} \boldsymbol{v}^i \cdot \boldsymbol{v}^i) \right] + \operatorname{div} \left[\varrho^i (\varepsilon^i + \frac{1}{2} \boldsymbol{v}^i \cdot \boldsymbol{v}^i) \boldsymbol{v}^i \right] = \\ &= \operatorname{div}(\boldsymbol{\sigma}^i \boldsymbol{v}^i - \boldsymbol{q}^i) + \varrho^i r^i + \varrho^i \boldsymbol{v}^i \cdot \boldsymbol{b}^i + \varepsilon^i_s + \\ &+ m^i (\varepsilon^i + \frac{1}{2} \boldsymbol{v}^i \cdot \boldsymbol{v}^i) + \boldsymbol{m}^i \cdot \boldsymbol{v}^i . \end{aligned} \tag{2-41}$$

The above equation can be simplified to yield

$$\begin{aligned}\frac{\partial}{\partial t}(\varrho^i\varepsilon^i) + \operatorname{div}(\varrho^i\varepsilon^i\boldsymbol{v}) = &-\operatorname{div}\boldsymbol{q}^i + \operatorname{tr}(\boldsymbol{\sigma}^i\boldsymbol{L}^i) + \\ &+ \varrho^i r^i + \varepsilon^i_s + m^i\varepsilon^i .\end{aligned} \tag{2-42}$$

The above equation can be further simplified by introducing

$$\varrho\varepsilon = \sum \varrho^i\varepsilon^i, \quad \boldsymbol{q} = \sum \boldsymbol{q}^i, \quad \boldsymbol{h} = \boldsymbol{q} + \sum \varrho^i\varepsilon^i\boldsymbol{u}^i, \quad \varrho r = \sum \varrho^i r^i, \tag{2-43}$$

and assuming

$$\sum \left[\varepsilon^i_s + m^i(\varepsilon^i + \frac{1}{2}\boldsymbol{v}^i\cdot\boldsymbol{v}^i) + \boldsymbol{m}^i\cdot\boldsymbol{v}^i\right] = 0, \tag{2-44}$$

and summing over i:

$$\varrho\dot{\varepsilon} = \sum \left[\operatorname{tr}(\boldsymbol{\sigma}^i\boldsymbol{L}^i) - \boldsymbol{m}^i\cdot\boldsymbol{v}^i - \frac{1}{2}m^i\boldsymbol{v}^i\cdot\boldsymbol{v}^i\right] - \operatorname{div}\boldsymbol{h} + \varrho r, \tag{2-45}$$

which governs the time rate at which the internal energy of the mixture is changing.

Second Law of Thermodynamics

The precise form of the second law of thermodynamics is the subject of much debate even within the frame-work of single constituents, and thus more so when we wander into the unchatered waters of materials with memory, and for processes that stray far from equilibrium. Also contentious are issues that pertain to whether such a law holds for a body or for all its arbitrary subparts, i.e., whether the law should only be posited in a global form or whether we can deduce local forms for them based on the arbitrariness of the subbodies. It is beyond the scope of these lectures to get into such issues.

There has been considerable success in the interpretation of the Clausius-Duhem inequality as the second law of thermodynamics within the context of the mechanics of a single continuum. While this does not provide a carte blanche for its use in mixtures, it definitely seems reasonable to determine the consequence of such an assumption and we shall do this here. The first question that confronts us is whether the second law holds for each constituent of the mixture or only to the mixture as a whole, and the only way to resolve this question is to obtain the consequences of the different options and determine which, if any, is the most appropriate.

Let η^i and T^i denote the specific entropy and absolute temperature of the ith constituent, respectively. Then, the global form of the Clausius-Duhem inequality takes the form

$$\frac{\partial}{\partial t}\int_{\mathcal{P}} \sum \varrho^i\eta^i dv + \int_{\partial\mathcal{P}} \sum \varrho^i\eta^i\boldsymbol{v}^i\cdot d\boldsymbol{a} \geq -\int_{\partial\mathcal{P}} \sum \frac{\boldsymbol{q}^i}{T^i}\cdot d\boldsymbol{a} + \int_{\mathcal{P}} \sum \frac{\varrho^i r^i}{T^i} dv, \tag{2-46}$$

which leads to the local form

$$\frac{\partial}{\partial t}\left(\sum \varrho^i\eta^i\right) + \operatorname{div}\left(\sum \varrho^i\eta^i\boldsymbol{v}\right) + \operatorname{div}\left(\sum \frac{\boldsymbol{q}^i + \varrho^i\eta^i T^i\boldsymbol{u}^i}{T^i}\right) - \sum \frac{\varrho^i r^i}{T^i} \geq 0\,. \tag{2-47}$$

In general the temperature field associated with the different constituents are different. However, there are many applications wherein we could assume that the temperatures of the all the constituents are the same, i.e., $T^i = T$. In this case, it can be shown that the mixture as a whole satisfies

$$\begin{aligned} &-\varrho(\dot{A} + \eta\dot{T}) + \sum\left(\operatorname{tr}\left[(\boldsymbol{\sigma}^i - \varrho^i(A^i - A)\mathbf{1})\,\boldsymbol{L}^i\right]\right) - \\ &-\frac{1}{2}m^i\boldsymbol{v}^i\cdot\boldsymbol{v}^i - \left[\boldsymbol{m}^i + \nabla(\varrho^i(A^i - A))\right]\cdot\boldsymbol{v}^i - \\ &-\frac{(\boldsymbol{q}^i + \varrho^i T\eta^i\boldsymbol{u}^i)\cdot\nabla T}{T} \geq 0\,, \end{aligned} \tag{2-48}$$

where

$$\eta := \frac{\sum \varrho^i\eta^i}{\varrho}\,, \quad A = \frac{\sum \varrho^i A^i}{\varrho}\,. \tag{2-49}$$

The inequality (2-48) is the local form of the Clausius-Duhem inequality for the mixture as a whole.

Volume Additivity Constraint

In the mechanics of a single continuum a constraint that is often appealed to is that of incompressibility. It is possible that one deals with a mixture of constituents which are individually incompressible. Such a possibility leads to very subtle issues. Consider for example a mixture of incompressible fluids, each of unit volume. When they are mixed together, it is plausible that the mixture is two units of volume. In reality, at any point in the actual mixture, we would have only one constituent. However, mixture theory presumes that at each point in the mixture there exists a particle belonging to each constituent, and this would imply that the volume of each constituent has doubled as the volume of the mixture is two units, thereby violating our presumption that the constituent was incompressible. This apparent dichotomy can however be resolved if we define equivalent homogenized reference configurations for the constituents of two units and then allowing them to co-exist in the domain of the mixture. We can now enforce the constraint of volume additivity for the constituents. For the sake of simplicity let us suppose that we have a mixture of two constituents, then (see Mills [23]) the volume additivity constraint takes the form

$$\frac{\varrho^1}{\varrho^1_R} + \frac{\varrho^2}{\varrho^2_R} = 1\,, \tag{2-50}$$

or equivalently

$$\frac{\varrho^1}{\varrho^1_R}(\operatorname{div} \boldsymbol{v}') + \frac{\varrho^2}{\varrho^2_R}(\operatorname{div} \boldsymbol{v}^2) + \nabla\left(\frac{\varrho^1}{\varrho^1_R}\right)\cdot(\boldsymbol{v}^1 - \boldsymbol{v}^2) = 0\,. \tag{2-51}$$

Other Constraints

In addition to the volume additivity constraint, numerous other constraints have been studied. While most internal constraints that are considered fall into the category of constraints that do no work, Green, Nagdhi and Trapp [24] studied internal constraints that produce no entropy. Constraints for mixtures are discussed in detail by Liu [25] and Hutter and Svendsen [26].

3. Constitutive Relations

In the previous section we have summarized the basic balance laws for mixtures. However, the system of equations is incomplete in that there are far more unknowns than equations and in order the render the equations determinate we have to provide constitutive relations for m^i, $\boldsymbol{\sigma}^i$, $\boldsymbol{m}^i$, $\boldsymbol{M}^i$, ε^i, $\boldsymbol{q}^i$, ε^i_s and η^i. For the remainder of our discussions we shall ignore m^i, $\boldsymbol{M}^i$, and ε^i_s, and for simplicity and clarity of analysis we shall restrict ourselves to just two constituents. The structure of the constitutive quantities would depend on the nature of the constituents, whether they are solid, liquid or gas, and futhermore on the specific nature of the solid, liquid or gas.

Elastic Solids Infused With A Fluid

First, let us consider the case of a mixture of a rubber-like solid and a linearly viscous liquid. Then, it may be reasonable to assume that (cf. Shi, Rajagopal and Wineman [27])

$$(A^s, A^f, A) = (A^s, A^f, A)(\varrho^f, \boldsymbol{F}^s, \boldsymbol{v}^s - \boldsymbol{v}^f)\,, \tag{3-1}$$

$$(\boldsymbol{\sigma}^s, \boldsymbol{\sigma}^f, \boldsymbol{m}) = (\boldsymbol{\sigma}^s, \boldsymbol{\sigma}^f, \boldsymbol{m})(\varrho^f, \boldsymbol{F}^s, \nabla\boldsymbol{F}^s, \boldsymbol{v}^s - \boldsymbol{v}^f, \boldsymbol{L}^s, \boldsymbol{L}^f, \boldsymbol{a}^{sf})\,, \tag{3-2}$$

where the superscript s and f refer to a solid and fluid, respectively, and

$$\boldsymbol{a}^{sf} = \frac{D^{(f)}\boldsymbol{v}^s}{Dt} - \frac{D^{(s)}\boldsymbol{v}^f}{Dt}\,. \tag{3-3}$$

The above measure of the relative acceleration between the constituents is frame-indifferent. A special structure for A that has led to reasonable results (see Rajagopal and Tao [1]) is

$$A = A_0(\varrho^f, \boldsymbol{F}^s) + \frac{1}{2}\boldsymbol{H}(\varrho^f, \boldsymbol{F}^s)(\boldsymbol{v}^s - \boldsymbol{v}^f)\cdot(\boldsymbol{v}^s - \boldsymbol{v}^f)\,, \tag{3-4}$$

where $\boldsymbol{H}$ is a symmetric, positive semi-definite tensor called the virtual mass tensor. If the elastic solid under consideration is isotropic, then it follows that

$$A_0(\varrho^f, \boldsymbol{F}^s) = \tilde{A}_0(\varrho^f, I_1, I_2, I_3), \tag{3-5}$$

and if the mixture satisfies the volume additivity constraint, then

$$A_0(\varrho^f, \boldsymbol{F}^s) = \hat{A}_0(\varrho^f, I_1, I_2), \tag{3-6}$$

where I_1, I_2 and I_3 are the principal invariants of $\boldsymbol{B}^s$.

If the solid under consideration is anisotropic, then A_0 can be appropriately modified. Thus, if the solid is orthotropic it would be reasonable to assume that

$$A_0(\varrho^f, \boldsymbol{F}^s) = A_0(\varrho^f, I_1, I_2, E_{11}, E_{22}, E_{33}, (E_{12})^2, (E_{13})^2, (E_{23})^2), \tag{3-7}$$

while if the solid material is transversely isotropic, then it would be natural to pick A_0 as

$$A_0(\varrho^f, \boldsymbol{F}^s) = A_0(\varrho^f, I_1, I_2, E_{33}, (E_{13})^2 + (E_{23})^2), \tag{3-8}$$

where we have assumed that the mixtures meet volume additivity.

For problems involving the diffusion of solvents like tolouene in rubber like elastic solids, based on the work of Wall and Flory [28], Shi, Rajagopal and Wineman [27] assumed that

$$\boldsymbol{H} = \boldsymbol{0}, \tag{3-9}$$

and

$$A_0 = K\left[I_1 - 3 + \left(1 - \frac{\varrho^f}{\varrho^f_R}\right)\right], \tag{3-10}$$

where $K = \frac{RT}{2M_c}$, R being the gas constant, T the absolute temperature and M_c the molecular weight between the cross-links of the solid. Gandhi, Rajagopal and Wineman [29] used another model for A_0, namely

$$A_0 = K(I_1 - 3) + \frac{RT}{V_1}\left[\frac{1-\nu}{\nu}(1-\nu) + \chi(1-\nu)\right], \tag{3-11}$$

where V_1 is the molar volume and χ is a mixing parameter which depends on the particular solid-fluid combination and ν is the volume fraction given by

$$\nu = \frac{\varrho^s}{\varrho^s_R}. \tag{3-12}$$

Dai and Rajagopal [30] studied the diffusion of a fluid through a transversely isotropic solid under the assumption that

$$\boldsymbol{H} = \boldsymbol{0}, \tag{3-13}$$

and

$$A_0 = K_1(I_1 - 3) + K_2 E_{33} + K_3(I_2 - 3) + K_4(I_1 - 3)^2 + \\ + K_5 E_{33}^2 + K_6(E_{13}^2 + E_{23}^2) + \alpha_1 \frac{\varrho^f}{\varrho_R^f} + \alpha_2 \left(\frac{\varrho^f}{\varrho_R^f} \right)^2 , \tag{3-14}$$

when the K_i^s, $i = 1, \ldots, s$, α_1 and α_2 are constants.

We shall not discuss the resolution of specific boundary value problems within the context of this theory, the details of the same can be found in [1].

Suspensions And Slurries

In the previous sub-section we considered a mixture of a solid and a fluid wherein the fluid is absorbed by a porous solid that may or may not be capable of swelling. Another class of mixtures that merit assiduous study are suspensions or slurries, namely mixtures wherein solid particles are suspended in a fluid. To warrant their modelling as mixtures it is imperative that the particles are sufficiently small and their volume fraction in the mixture reasonably large. A theory that is capable of describing the turbulent motion of suspensions is attempted in [1], however here we shall rest content with a simplified description of suspensions. Suspensions have been modeled in a variety of ways (Einstein [31], [32], [33], Jeffrey [34], Taylor [35], Guth and Simha [36], Brenner [37], Batchelor [38]), here we shall discuss the modelling of suspensions as a mixture of a granular particulate media and a fluid. We shall restrict ourselves to a purely mechanical mixture and ignore thermal effects. In many of the modelling of suspensions as a two-constituent* mixture, the solid particles are also assumed to behave like a linearly viscous fluid (see Murray [39], Anderson and Jackson [40], El-Kaissy [41]) with a viscosity μ^s and an associated pressure p^s whose meaning is unclear and furthermore leads to an indeterminacy in the system of equations that is redressed by some adhoc relationship between the pressures in the fluid and solid.

We shall assume that the partial stresses in both the constituents and the interaction term depend on

$$\varrho^s \, , \varrho^f \, , \operatorname{grad} \varrho^s \, , \operatorname{grad} \varrho^f \, , \boldsymbol{v}^f - \boldsymbol{v}^s \, , \boldsymbol{D}^s \, , \boldsymbol{D}^f \, . \tag{3-15}$$

If we are interested in accommodating the influence of the accelerations and the spins of the constituents we would have to allow for the interaction term $\boldsymbol{I}$ to depend on a frame invariant measure of the relative acceleration $\boldsymbol{a}^{sf}$ and probably the relative spin $\boldsymbol{W}^s - \boldsymbol{W}^f$, where

$$\boldsymbol{a}^{sf} = \frac{D^{(s)} \boldsymbol{v}^f}{Dt} - \frac{D^{(f)} \boldsymbol{v}^s}{Dt} \, . \tag{3-16}$$

---

*We invariably come across the terminology two-phase instead of two-constituent, in all papers devoted to this subject. As the granular material and the fluid are not two-phases of the same substance I shall refrain from this popularly accepted terminology.

We shall simplify our modelling further by assuming that the partial stress in each constituent depends only on the density and kinematical quantities associated with that constituent**, i.e.,

$$\boldsymbol{\sigma}^s = \boldsymbol{\sigma}^s(\varrho^s, \operatorname{grad}\varrho^s, \boldsymbol{D}^s)\,, \tag{3-17}$$

$$\boldsymbol{\sigma}^f = \boldsymbol{\sigma}^f(\varrho^f, \operatorname{grad}\varrho^f, \boldsymbol{D}^f)\,, \tag{3-18}$$

while

$$\boldsymbol{m} = \boldsymbol{m}(\varrho^s, \varrho^f, \boldsymbol{v}^s - \boldsymbol{v}^f)\,, \tag{3-19}$$

that is the interaction between the two constituents depends on physical quantities associated with both the constituents.

The dependance of $\boldsymbol{T}^f$ on the grad ϱ^f would allow for the model to include the Korteweg fluid as a special case. However, as we are merely interested in outlining a simple model, we shall make the additional assumption that

$$\boldsymbol{\sigma}^f \doteq \boldsymbol{\sigma}^f(\varrho^f, \boldsymbol{D}^f) = -p^f\mathbf{1} + \lambda^f(\varrho^f)(\operatorname{tr}\boldsymbol{D}^f)\mathbf{1} + 2\mu^f(\varrho^f)\boldsymbol{D}^f\,, \tag{3-20}$$

namely that the partial stress for the fluid is given by the classical linearly viscous model. It is customary to express $\boldsymbol{\sigma}^f$ as

$$\boldsymbol{\sigma}^f = (1-\nu)\left\{-\tilde{p}^f\mathbf{1} + \tilde{\lambda}^f(\varrho^f)(\operatorname{tr}\boldsymbol{D}^f)\mathbf{1} + 2\tilde{\mu}^f(\varrho^f)\boldsymbol{D}^f\right\} \tag{3-21}$$

where

$$(1-\nu) \equiv \frac{\varrho^f}{\varrho^f_R}\,. \tag{3-22}$$

Using standard arguments in continuum mechanics, we can obtain representations for the partial stress $\boldsymbol{T}^s$. Here, we shall merely assume that (see [45])

$$\begin{aligned}\boldsymbol{\sigma}^s = \{\beta_0(\varrho^s) + \beta_1(\varrho^s)\operatorname{grad}\varrho^s\cdot\operatorname{grad}\varrho^s + \beta_2(\varrho^s)\operatorname{tr}\boldsymbol{D}^s\}\,\mathbf{1} + \\ + \beta_3(\varrho^s)\operatorname{grad}\varrho^s\otimes\operatorname{grad}\varrho^s + \beta_4(\varrho^s)\boldsymbol{D}^s\,.\end{aligned} \tag{3-23}$$

In the above expression, the terms within the curly brackets can be viewed as the term corresponding to p^s and thus the model resembles a fluid with an equation of state for the term p^s given as above.

It should be emphasized that the term grad ϱ^s introduces difficulties with regard to the specification of boundary conditions.

For a large class of problems an interaction term of the form

$$\boldsymbol{m} = \alpha_1\operatorname{grad}\varrho^s + \alpha_2\operatorname{grad}\varrho^f + \alpha_3(\boldsymbol{v}^s - \boldsymbol{v}^f) + \alpha_4\boldsymbol{a}^{sf} \tag{3-24}$$

**This assumption is usually given the title "Principle of phase separation" (see Adkins [42], [43], Drew [44] and Rajagopal, Massoudi and Ekmann [45]).

would suffice, though we can include terms to incorporate Magnus effect, Faxen forces and Basset forces.

Having made the above constitutive assumptions, we can obtain the relevant governing equations by substituting the above in the balance of linear momentum for each constituent. Using the above frame-work Johnson, Massoudi and Rajagopal [46], [47], [48] have studied the flow of a fluid infused with particles in various geometries. The predictions of the theory seem to be in keeping with physics, and in the few cases where comparison with experiments are possible, it provides reasonable concordance.

We would like to caution the reader that care has to exercised in applying mixture theory for modelling fluids infused with solid particles, especially if the volume fraction of the solids is low. It should also be borne in mind that the above theory does not take into account the distribution of both sizes and shapes of the particles. The above theory might be appropriate for a mixture of a fluid with nearly spherical particles that are not disparately sized, in fact the particles have to be reasonably small in terms of some length scale natural to the problem if mixture theory is to be deemed relevant.

Mixture of Two Fluids

There have been many studies concerning the mixture of two fluids within the context of mixture theory (see Adkins [42], Craine [49], Green and Nagdhi [50], Mills [51], Bowen [52], Müller [53], [54], Williams and Sampaio [55]). In [1], Rajagopal and Tao appeal to an idea due to Soo [56] to model a mixture of two immiscible incompressible Newtonian fluids that is very different from the modelling that is presented here. Here, we provide a simpler and direct modelling of the mixture, albeit its range of applicability might be somewhat restrictive. Once again, we shall restrict ourselves to a purely mechanical frame-work, an extension to a thermodynamic setting being straightforward.

Based on our understanding of Newtonian fluids and mixtures, we would expect the partial stresses in the two fluids to depend on the density of the two fluids and the symmetric part of their respective velocity gradients, i.e.,

$$\boldsymbol{\sigma}^{(1)} = \boldsymbol{\sigma}^{(1)}(\varrho^{(1)}, \varrho^{(2)}, \boldsymbol{D}^{(1)}, \boldsymbol{D}^{(2)}), \tag{3-24}$$

$$\boldsymbol{\sigma}^{(2)} = \boldsymbol{\sigma}^{(2)}(\varrho^{(1)}, \varrho^{(2)}, \boldsymbol{D}^{(1)}, \boldsymbol{D}^{(2)}), \tag{3-25}$$

where superscripts (1) and (2) denote the two components. At this stage we could make the additional assumption of "phase separation", however we shall not do so as there is considerable evidence within the context of a mixture of two fluids the partial stresses of both the constituents seem to depend on the kinematical variables associated with both of them. We could include as additional dependant variables the relative velocity $\boldsymbol{v}^{(1)} - \boldsymbol{v}^{(2)}$, the relative spin $\boldsymbol{W}^{(1)} - \boldsymbol{W}^{(2)}$ and the frame invariant measure of the relative acceleration $\boldsymbol{a}^{(12)}$, but in the interest of simplicity of illustration we shall leave it out.

We shall assume that the interactive force is given by

$$\boldsymbol{m} = \boldsymbol{m}(\varrho^{(1)}, \varrho^{(2)}, \boldsymbol{v}^{(1)} - \boldsymbol{v}^{(2)}, \boldsymbol{W}^{(1)} - \boldsymbol{W}^{(2)}, \boldsymbol{a}^{(12)}) . \tag{3-26}$$

A simple representation for $\boldsymbol{\sigma}^{(1)}$ and $\boldsymbol{\sigma}^{(2)}$ is

$$\begin{aligned} \boldsymbol{\sigma}^{(1)} &= \left[-p_1(\varrho^{(1)}) + \lambda_{11}(\varrho^{(1)}) \operatorname{tr} \boldsymbol{D}^{(1)} + \lambda_{12}(\varrho^{(2)}) \operatorname{tr} \boldsymbol{D}^{(2)}\right] 1 + \\ &\quad + 2\mu_{11}(\varrho^{(1)}) \boldsymbol{D}^{(1)} + 2\mu_{12}(\varrho^{(2)}) \boldsymbol{D}^{(2)} , \end{aligned} \tag{3-27}$$

$$\begin{aligned} \boldsymbol{\sigma}^{(2)} &= \left[-p_2(\varrho^{(2)}) + \lambda_{21}(\varrho^{(1)}) \operatorname{tr} \boldsymbol{D}^{(1)} + \lambda_{22}(\varrho^{(2)}) \operatorname{tr} \boldsymbol{D}^{(2)}\right] 1 + \\ &\quad + 2\mu_{21}(\varrho^{(1)}) \boldsymbol{D}^{(1)} + 2\mu_{22}(\varrho^{(2)}) \boldsymbol{D}^{(2)} . \end{aligned} \tag{3-28}$$

Of course, the above model is quite ad hoc but allows for the effects of cross-viscosities μ_{12} and μ_{21} to play a role in the modelling.

Mixture of two liquids, and a liquid and a gas have numerous interesting applications. The flow of oil and water emulsions in geometries that are relevant to lubrication problem have been studied by Chamniprasart et-al [57], Wang et-al [58] and Al-arabi et-al [59].

4. Boundary Conditions

The greatest impediment to putting the continuum theory of mixtures to use is the lack of a clear understanding of the boundary conditions that ought to be applied at the boundary of a mixture. In general, we have information regarding the traction, displacement or velocity at each point on the boundary. As mixture theory requires that each point in space be occupied by a particle belonging to each constituent, we need to have an unambiguous method for determing how the information on the boundary can be assigned to the various constituents. Thus, for instance in traction boundary value problems, we know the total traction on the boundary but not the individual particial tractions. At first glance, we might think that this situation is peculiar to traction boundary value problems, and does not raise its ugly head in boundary value problems where displacement or velocity are to be prescribed. However, a moment's thought makes it obvious that even in displacement (or velocity) boundary value problems, all that is known is information at a point on the boundary of a mixture. It might seem "natural" to assign, for instance, in the case of a mixture of two fluids flowing in a pipe the "no-slip" boundary condition for each component; however, this is an additional assumption for "no-slip" for the mixture as a whole does not imply the same for the individual constituents and even in the case of the flow of fluid mixtures there are many situations wherein it is not at all clear what the pertinent boundary conditions for the constituents are. In fact, the "no-slip" boundary condition is an assumption that seems to be reasonable for a large class of problems involving viscous fluids. However, even in the case of single constituent fluids it is far from clear that such a condition prevails when dealing with thin films or at the contact of free surfaces adjacent to moving boundaries (dynamic contact associated with wetting).

Thus, it is not surprising that not many problems have been solved within the context of the theory of mixtures. Most of the early attempts at applying the theory of mixtures were devoted to the propagation of waves in infinite bodies of mixtures, thereby obviating the need for specifying boundary conditions. Traction boundary value problems are the most challenging with regard to the resolution of the indeterminacy discussed earlier, and it is this issue that we shall address here.

For a special class of mixtures, namely a solid infused with a fluid, Gandhi, Rajagopal and Wineman [29] generalized the earlier work of Shi, Wineman and Rajagopal [27] and obtained additional conditions on the boundary.

Let us consider a mixture that consists of a fluid infused in a solid and let us focus our attention at any point belonging to the boundary of the mixture. Next suppose that we have an unit cube of the mixture of solid which after absorbing the liquid is now swollen such that it is homogeneously deformed to the same state of deformation as the point of interest on the boundary, that is each point in the swollen cube is in the same state of deformation as well as the densities of the two constituents as that at the point on the boundary. If such a cube is saturated, then we could require that the variation in the Gibbs free energy of dilution is zero, and this provides a relationship between the extent of deformation of the mixture, the densities and the applied tractions. This in precisely the manner in which the Flory-Huggins relation is derived in the theory of rubber elasticity, though there it is applied to the actual swollen cube, rather than our thought experiment here that has been developed with a view towards providing boundary conditions.

There is no point getting into an extensive detailed discussion of how the above boundary conditions are generated, here. The interested reader can find the details in the recent book by Rajagopal and Tao [1], and the paper in which these ideas were first given form (Gandhi, Rajagopal and Wineman [29]). However, it might be worthwhile to highlight the deficiencies of the method, the most telling of which is the consideration of a thermodynamically open system to be a closed one. The variational problem that is considered is not that for the Gibbs's free energy for dilution but rather one for the specific Helmholtz potential; in fact the variation in the specific Helmholtz potential is equated to the variation in the work due to the forces acting on the swollen cube. However, fluid is allowed to enter and leave the cube and hence the system under consideration is open.

Also, the boundary conditions generated on the basis of the above arguments only apply for a very special class of problems, namely those for which the boundary is saturated, a condition that is not relevant to many technical problems.

Being dissatisfied with the above boundary condition, Rajagopal and Tao [1] have recently proposed another that has purely mechanical underpinnings. They assume that the total traction is split into the partial tractions in the ratio of the specific densities of the constituents, that is the traction is weighted according to the specific densities. Using the above assumption, Rajagopal and Tao [1] restudy the diffusion of a fluid through a solid capable of absorbing the fluid, a problem that was initially studied by Shi, Rajagopal and Wineman [27]*.

A natural additional boundary condition for problems involving the diffusion of fluids through a solid is the condition that stems from requiring that the chemical potential be continuous across the boundary between the mixture and the pure fluid. This condition has not been investigated and deserves looking into.

We shall document without derivation the additional conditions that stem from the notion that the boundary is saturated, the interested reader can find details in [1]:

$$\boldsymbol{\sigma} = \left[\varrho_R^f + (\varrho_R^f - \varrho^f)\varrho\frac{\partial A}{\partial \varrho^f}\right]\mathbf{1} + 2\varrho\left[\left(\frac{\partial A}{\partial I_1} + \frac{\partial A}{\partial I_2}I_1\right)\boldsymbol{B} - \frac{\partial A}{\partial I_2}\boldsymbol{B}^2\right]. \tag{4-1}$$

The traction splitting assumption proposed later by Rajagopal and Tao [1] takes the form, in the case of just two constituents,

$$\boldsymbol{t}^1 = (\boldsymbol{\sigma}^1)^\top \boldsymbol{n} = \frac{\varrho^1}{\varrho_R^1}\boldsymbol{t}, \tag{4-2}$$

$$\boldsymbol{t}^2 = (\boldsymbol{\sigma}^2)^\top \boldsymbol{n} = \frac{\varrho^2}{\varrho_R^2}\boldsymbol{t}. \tag{4-3}$$

Notice that (4-2) and (4-3) automatically satisfy the requirements of the volume additivity constraint.

For problems involving a mixture of two fluids, it might be reasonable to assume that the velocities of both the constituents at a solid boundary are zero. However, in dealing with free surface problems involving a mixture of two fluids we are once again presented with a difficulty as only the total stress for the mixture is zero. Once again, as a first step, we may assume that the partial stresses are individually zero.

Boundary conditions for flows involving a fluid infused with solid particles are complicated by the fact that the particles can and do slip on the solid boundary. While it may be reasonable to assume the no-slip velocity condition for the fluid, this may or may not be appropriate for the solid. We refer the reader to [46], [47], and [48] where several boundary value problems have been studied for such mixtures.

*The study of Shi, Rajagopal and Wineman [27] suffers from a serious drawback in that the specific deformation field for the solid and the velocity field for the fluid leads to an inconsistency that cannot be resolved without appealing to rather convoluted artifices. This however does not impugn the boundary condition but the specific forms for the deformation and velocity fields assumed in that study. A detailed discussion of the relevant issues can be found in the book by Rajagopal and Tao [1]

5. Moving Singular Surfaces

Next, we shall discuss the problem wherein there is a moving singular surface, ahead and behind which there are mixtures with differing properties. A special case is that which corresponds to a single constituent ahead of the surface while there is a mixture behind the surface, a situation that has relevance to the spreading of contaminants. Here, we shall provide a very brief discussion of the method for generating the appropriate equations without getting into details, the interested reader is referred to [1] for a more complete treatment.

Consider the situation depicted in Figure 1 which assumes a body that can be thought of as three distinct regions; a transition region which separates two regions of mixtures.

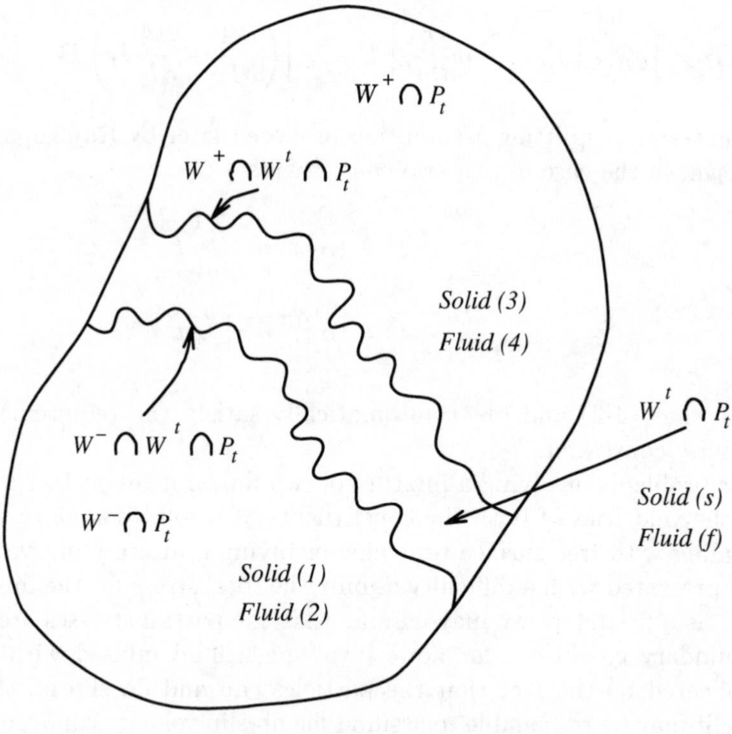

FIGURE 1.

The modelling assumes that the standard equations for a mixture outlined in section 3 apply on either side of the transitional layer, and presumes that in the limit the layer can be treated as a singular surface with which certain surface properties can be associated. A key feature of the modelling is the derivation of equations governing the propagation of the singular surface. These equations are obtained by considering a part of the body which includes parts of the mixture and the singular surface (see Fig. 1). As the derivations are lengthy and involved, we shall not get into them here, but refer the reader to [1] for the details of the same. Here, we shall just define the appropriate variables on the singular surface and then document the relevant field equations. Let us construct a co-ordinate system (l^1, l^2, l^3) such that the axes l^1 and l^2 lie on the tangent plane to S and the third axes l^3 is perpendicular to the plane.

Now, we introduce the surface properties for the fluid f and the solid s on S in the following manner:

(i) surface mass densities

$$\hat{\varrho}_f \equiv \int_\delta \tilde{\varrho}_f dl^3\,, \quad \hat{\varrho}_s \equiv \int_\delta \tilde{\varrho}_s dl^3\,; \tag{5-1}$$

(ii) surface velocities

$$\hat{v}_{fi} \equiv \frac{1}{\hat{\varrho}_f}\int_\delta \tilde{\varrho}_f \tilde{v}_{fi} dl^3\,, \quad \hat{v}_{si} \equiv \frac{1}{\hat{\varrho}_s}\int_\delta \tilde{\varrho}_s \tilde{v}_{si} dl^3\,; \tag{5-2}$$

(iii) surface body forces

$$\hat{b}_{fi} \equiv \frac{1}{\hat{\varrho}_f}\int_\delta \tilde{\varrho}_f \tilde{b}_{fi} dl^3\,, \quad \hat{b}_{si} \equiv \frac{1}{\hat{\varrho}_s}\int_\delta \tilde{\varrho}_s \tilde{b}_{si} dl^3\,; \tag{5-3}$$

(iv) surface momentum supplies

$$\hat{m}_{fi} \equiv \int_\delta \tilde{m}_{fi} dl^3\,, \quad \hat{m}_{si} \equiv \int_\delta \tilde{m}_{si} dl^3\,; \tag{5-4}$$

(v) surface stress tensors $(\hat{\sigma}_f^{i\alpha}, \hat{\sigma}_s^{i\alpha})$

$$\hat{\sigma}_f^{i\alpha} N_\alpha \equiv \int_\delta \tilde{\sigma}_f^{ji} n_j dl^3\,, \quad \hat{\sigma}_s^{i\alpha} N_\alpha \equiv \int_\delta \tilde{\sigma}_s^{ji} n_j dl^3\,; \tag{5-5}$$

where $\boldsymbol{N}$ is the unit vector on S tangent to S and normal to $\Gamma = S' \cap S$, $\alpha = 1, 2$ (Figure 2).

(vi) surface moment of momentum supplies

$$\hat{M}_{fi} \equiv \int_\delta \tilde{M}_{fi} dl^3\,, \quad \hat{M}_{si} \equiv \int_\delta \tilde{M}_{si} dl^3\,; \tag{5-6}$$

(vii) surface internal energy densities

$$\hat{\varepsilon}_f \equiv \frac{1}{\hat{\varrho}_f}\int_\delta \tilde{\varrho}_f \tilde{\varepsilon}_f dl^3\,, \quad \hat{\varepsilon}_s \equiv \frac{1}{\hat{\varrho}_s}\int_\delta \tilde{\varrho}_s \tilde{\varepsilon}_s dl^3\,; \tag{5-7}$$

(viii) surface heat fluxes $(\hat{q}_f^\alpha, \hat{q}_s^\alpha)$

$$\hat{q}_f^\alpha N_\alpha \equiv \int_\delta \tilde{q}_{fi} n_i dl^3 , \quad \hat{q}_s^\alpha N_\alpha \equiv \int_\delta \tilde{q}_{si} n_i dl^3 ; \tag{5-8}$$

(ix) surface external heat supplies

$$\hat{r}_f \equiv \frac{1}{\hat{\varrho}_f} \int_\delta \tilde{\varrho}_f \tilde{r}_f dl^3 , \quad \hat{r}_s \equiv \frac{1}{\hat{\varrho}_s} \int_\delta \tilde{\varrho}_s \tilde{r}_s dl^3 ; \tag{5-9}$$

(x) surface energy supplies

$$\hat{\varepsilon}_{sf} \equiv \int_\delta \tilde{\varepsilon}_{sf} dl^3 , \quad \hat{\varepsilon}_{ss} \equiv \int_\delta \tilde{\varepsilon}_{ss} dl^3 ; \tag{5-10}$$

(xi) specific surface entropies

$$\hat{\eta}_f \equiv \frac{1}{\hat{\varrho}_f} \int_\delta \tilde{\varrho}_f \tilde{\eta}_f dl^3 , \quad \hat{\eta}_s \equiv \frac{1}{\hat{\varrho}_s} \int_\delta \tilde{\varrho}_s \tilde{\eta}_s dl^3 ; \tag{5-11}$$

(xii) surface temperatures

$$\hat{T}_f \equiv \frac{1}{\hat{\varrho}_f} \int_\delta \tilde{\varrho}_f \tilde{T}_f dl^3 , \quad \hat{T}_s \equiv \frac{1}{\hat{\varrho}_s} \int_\delta \tilde{\varrho}_s \tilde{T}_s dl^3 ; \tag{5-12}$$

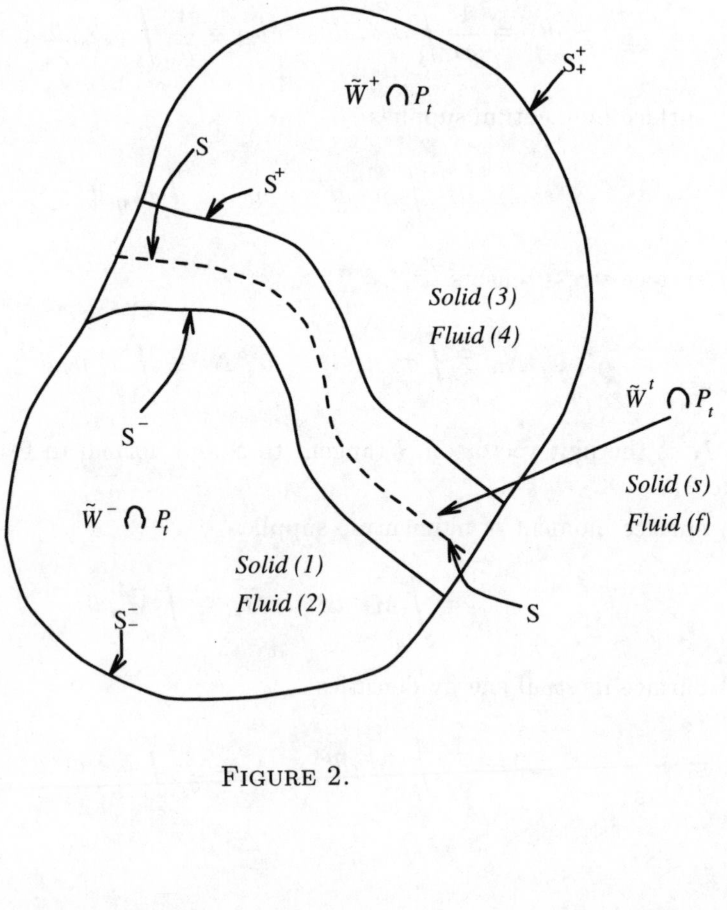

FIGURE 2.

The appropriate field equations for balance of mass, linear momentum, angular momentum, energy and the second law are:

$$\varrho_1(\hat{v}^i_f - \hat{v}^i_{(1)})n_i + \varrho_3(\hat{v}^i_{(3)} - \hat{v}^i_f)n_i + \frac{d\hat{\varrho}_s}{dt} + \\ + (a^{\alpha\beta}\hat{v}_{sj}t^j_\alpha)_{,\beta}\hat{\varrho}_s - 2K_m\hat{\varrho}_s\hat{v}_{sn} = 0\,, \tag{5-13}$$

$$\varrho_2(\hat{v}^i_f - \hat{v}^i_{(2)})n_i + \varrho_4(\hat{v}^i_{(4)} - \hat{v}^i_f)n_i + \frac{d\hat{\varrho}_f}{dt} + \\ + (a^{\alpha\beta}\hat{v}_{fj}t^j_\alpha)_{,\beta}\hat{\varrho}_f - 2K_m\hat{\varrho}_f\hat{v}_{fn} = 0\,, \tag{5-14}$$

$$\varrho_2 v^i_{(2)}(\hat{v}^j_f - v^j_{(2)})n_j + \varrho_4(v^j_{(4)} - \hat{v}^j_f)n_j \\ + \frac{d}{dt}(\hat{\varrho}_f\hat{v}^i_f) + (a^{\alpha\beta}\hat{v}_{fj}t^j_\alpha)_{,\beta}\hat{\varrho}_f\hat{v}^i_f - 2K_m\hat{\varrho}_f\hat{v}^i_f\hat{v}_{fn} \\ = \hat{\varrho}_f\hat{b}^i_f + \hat{m}^i_f + (-\sigma^{ji}_{(2)} + \sigma^{ji}_{(4)})n_j + (\hat{\sigma}^{i\alpha}_f)_{,\alpha}\,, \tag{5-15}$$

$$\varrho_1 v^i_{(1)}(\hat{v}^j_f - v^j_{(1)})n_j + \varrho_3(v^j_{(3)} - \hat{v}^j_f)n_j \\ + \frac{d}{dt}(\hat{\varrho}_s\hat{v}^i_s) + (a^{\alpha\beta}\hat{v}_{sj}t^j_\alpha)_{,\beta}\hat{\varrho}_s\hat{v}^i_s - 2K_m\hat{\varrho}_s\hat{v}^i_s\hat{v}_{sn} \\ = \hat{\varrho}_s\hat{b}^i_s + \hat{m}^i_s + (-\sigma^{ji}_{(1)} + \sigma^{ji}_{(3)})n_j + (\hat{\sigma}^{i\alpha}_s)_{,\alpha}\,, \tag{5-16}$$

$$\hat{m}^i_f + \hat{m}^i_s = 0\,, \tag{5-17}$$

$$\varepsilon_{ijk}t^j_\alpha\hat{\sigma}^{k\alpha}_f + \hat{M}_{fi} = 0\,, \tag{5-18}$$

$$\varepsilon_{ijk}t^j_\alpha\hat{\sigma}^{k\alpha}_f + \hat{M}_{si} = 0\,, \tag{5-19}$$

$$\varrho_2(\varepsilon_2 + \frac{1}{2}v^i_{(2)}v_{(2)i})(\hat{v}^j_f - v^j_{(2)})n_j \\ + \varrho_4(\varepsilon_4 + \frac{1}{2}v^i_{(4)}v_{(4)i})(\hat{v}^j_{(4)} - v^j_f)n_j + \frac{d}{dt}\left[\hat{\varrho}_f\left(\hat{\varepsilon}_f + \frac{1}{2}\hat{v}^i_f\hat{v}_{fi}\right)\right] \\ + (a^{\alpha\beta}\hat{v}_{fj}t^j_\alpha)_{,\alpha}\hat{\varrho}_f\left(\hat{\varepsilon}_f + \frac{1}{2}\hat{v}^i_f\hat{v}_{fi}\right) - 2K_m\hat{\varrho}_f\left(\hat{\varepsilon}_f + \frac{1}{2}\hat{v}^i_f\hat{v}_{fi}\right) \\ = \hat{\varrho}_f\hat{b}^i_f\hat{v}_{fi} + \hat{m}^i_f\hat{v}_{fi} + (-\sigma^{ji}_{(2)}v_{(2)i} + \sigma^{ji}_{(4)}v_{(4)i})n_j \\ + (q^i_{(2)} - q^i_{(4)})n_i + \hat{\varrho}_f\hat{r}_f + \hat{\varepsilon}_{sf} + (\hat{\sigma}^{i\alpha}_f\hat{v}_{fi} - \hat{q}^\alpha_f)_{,\alpha}\,, \tag{5-20}$$

$$
\begin{aligned}
&\varrho_1(\varepsilon_1 + \frac{1}{2}v^i_{(1)}v_{(1)i})(\hat{v}^j_f - v^j_{(1)})n_j \\
&\quad + \varrho_3(\varepsilon_3 + \frac{1}{2}v^i_{(3)}v_{(3)i})(\hat{v}^j_{(3)} - v^j_f)n_j + \frac{d}{dt}\left[\hat{\varrho}_s\left(\hat{\varepsilon}_s + \frac{1}{2}\hat{v}^i_s\hat{v}_{si}\right)\right] + \\
&\quad + (a^{\alpha\beta}\hat{v}_{sj}t^j_\alpha)_{,\alpha}\hat{\varrho}_s\left(\hat{\varepsilon}_s + \frac{1}{2}\hat{v}^i_s\hat{v}_{si}\right) - 2K_m\hat{\varrho}_s\left(\hat{\varepsilon}_s + \frac{1}{2}\hat{v}^i_s\hat{v}_{si}\right)\hat{v}_{sn} \\
&\quad = \hat{\varrho}_s\hat{b}^i_s\hat{v}_{si} + \hat{m}^i_s\hat{v}_{si} + (-\sigma^{ji}_{(1)}v_{(1)i} + \sigma^{ji}_{(3)}v_{(3)i})n_j + \\
&\quad + (q^i_{(1)} - q^i_{(3)})n_i + \hat{\varrho}_s\hat{r}_s + \hat{\varepsilon}_{ss} + (\hat{\sigma}^{i\alpha}_s\hat{v}_{si} - \hat{q}^\alpha_s)_{,\alpha}\,,
\end{aligned}
\tag{5-21}
$$

$$
\hat{m}^i_f\hat{v}_{fi} + \hat{m}^i_s\hat{v}_{si} + \hat{\varepsilon}_{sf} + \hat{\varepsilon}_{ss} = 0\,, \tag{5-22}
$$

and

$$
\begin{aligned}
&\varrho_1\eta_1(\hat{v}^j_f - v^j_{(1)})n_j + \varrho_2\eta_2(\hat{v}^j_f - v^j_{(2)})n_j + \\
&\quad + \varrho_3\eta_3(\hat{v}^j_{(3)} - v^j_f)n_j + \varrho_4\eta_4(\hat{v}^j_{(4)} - v^j_f)n_j + \\
&\quad + \frac{d}{dt}(\hat{\varrho}_f\hat{\eta}_f) + (a^{\alpha\beta}\hat{v}_{fj}t^j_\alpha)_{,\beta}\hat{\varrho}_f\hat{\eta}_f - 2K_m\hat{\varrho}_f\hat{\eta}_f\hat{v}_{fn} + \\
&\quad + \frac{d}{dt}(\hat{\varrho}_s\hat{\eta}_s) + (a^{\alpha\beta}\hat{v}_{sj}t^j_\alpha)_{,\beta}\hat{\varrho}_s\hat{\eta}_s - 2K_m\hat{\varrho}_s\hat{\eta}_s\hat{v}_{sn} \geq \\
&\quad \geq \left(\frac{q^i_{(1)}}{T_1} + \frac{q^i_{(2)}}{T_2} - \frac{q^i_{(3)}}{T_3} - \frac{q^i_{(4)}}{T_4}\right)n_i + \frac{\hat{\varrho}_f\hat{r}_f}{\hat{T}_f} + \frac{\hat{\varrho}_s\hat{r}_s}{\hat{T}_s} - \\
&\quad - \left(\frac{\hat{q}^\alpha_f}{\hat{T}_f} + \frac{\hat{q}^\alpha_s}{\hat{T}_s}\right)_{,\alpha}.
\end{aligned}
\tag{5-23}
$$

In the above equations (see Eisenhart [60], Kosinski [61])

$$
a_{\alpha\beta} = t^i_\alpha t^i_\beta\,, \tag{5-24}
$$

$$
t^i_\alpha = \frac{\partial x^i}{\partial l^\alpha}\,, \tag{5-25}
$$

$$
K_m = \frac{1}{2}a^{\alpha\beta}b_{\alpha\beta}\,, \tag{5-26}
$$

$$
b_{\alpha\beta} = n_i\frac{\partial^2 x^i}{\partial l^\alpha \partial l^\beta}\,. \tag{5-27}
$$

In the region ahead of and behind the singular surface, the equations for mixtures (the equations (5-13)–(5-23)) are supposed to hold. In order to solve the equations on S it is necessary to specify constitutive relations for $\hat{m}^i_f$, $\hat{m}^i_s$, $\hat{\boldsymbol{\sigma}}^i_f$, $\hat{\boldsymbol{\sigma}}^i_s$, $\hat{\boldsymbol{M}}_{fi}$, $\hat{\boldsymbol{M}}_{si}$, $\hat{\varepsilon}_f$, $\hat{\varepsilon}_s$, $\hat{\boldsymbol{q}}^\alpha_f$, $\hat{\boldsymbol{q}}^\alpha_s$, $\hat{\varepsilon}_{sf}$, $\hat{\varepsilon}_{ss}$, $\hat{\eta}_f$, and $\hat{\eta}_s$. Within the framework of the above theory, the movement of a singular surface in a mixture, within the content of different geometries, has been solved (see [1]) and the predictions of the theory seem to be in keeping with physical expectation.

6. Darcy's Equation

The seminal work of Fick [11] and Darcy [18] governing the diffusion of a constituent through one or more constituents has spawned various generalizations that have proved tremendously useful, but as with most successful approximations, they have been abused by applying them beyond their domain of validity and most definitely the intent of the pioneers. In this section, I shall briefly outline the status of Darcy's equation within the context of the theory of mixtures. A discussion on the status of Fick's equation within the context of mixtures can be found in Truesdell [2], Bowen [4], Adkins [42] or Atkin and Craine [3].

Let us consider the diffusion of a fluid through a solid. The appropriate equation for the balance of linear momentum for the constituents is (2-31). These equations do not as yet reflect the nature of the solid or fluid, these being provided by the constitutive assumptions. Let us suppose that there are no chemical reactions, and that neither of the constituents are created or destroyed, i.e., $m^i = 0$. Further, let us suppose that the balance of linear momentum for the solid constituent can be ignored. We recall that Darcy's equation essentially relates the mass flux of the fluid to the pressure gradient and is completely silent about the manner in which the solid is deformed, justifying our utter disregard for the balance of linear momentum for the solid. We are thus left with the balance of linear momentum for the fluid that reads

$$\operatorname{div} \boldsymbol{\sigma}^f - \boldsymbol{m} + \varrho^f \boldsymbol{b} = \varrho^f \frac{d\boldsymbol{v}^f}{dt}. \tag{6-1}$$

Let us consider the flow of a fluid that is slow enough to render inertial terms inconsequential. Here, we hasten to add that there are numerous generalizations to Darcy's equation that incorporate the effects of inertia, however our interest is to investigate the assumptions that lead to Darcy's equation in its original simple form. Thus, we are left with

$$\operatorname{div} \boldsymbol{\sigma}^f - \boldsymbol{m} + \varrho^f \boldsymbol{b} = \boldsymbol{0}. \tag{6-2}$$

All that remains is to make appropriate constitutive assumptions for the interaction term $\boldsymbol{m}$ and the partial stress for the fluid $\boldsymbol{\sigma}^f$. We shall assume that the interaction term $\boldsymbol{m}$ is given by

$$\boldsymbol{m} = \alpha \boldsymbol{v}^f, \tag{6-3}$$

where α is a constant. The above assumption is equivalent to assuming that the only interaction is due to the relative velocity between the fluid and solid, the latter being assumed to be at rest; α is usually referred to as the drag coefficient and depends on the viscosity of the fluid and the porosity of the solid. Finally, we assume that the partial stress for the fluid $\boldsymbol{\sigma}^f$ is given by

$$\boldsymbol{\sigma}^f = -p\mathbf{1}. \tag{6-4}$$

The last two assumptions might seem to be at odds with one another in that the partial stress for the fluid resembles that for an ideal fluid while the expression for $\boldsymbol{m}$ implies that we have "drag", a consequence of viscosity that is glaringly absent in the expression for the partial stress for the fluid. The two assumptions (6-3) and (6-4) should be viewed together as describing the effect of the fluid in the diffusion process, with the understanding that the effect due to the viscosity of the fluid manifests itself though $\boldsymbol{m}$, and it should be clearly borne in mind that we are not dealing with an ideal fluid.

It follows from (6-2) through (6-4) that

$$\boldsymbol{v}^f = \frac{1}{\alpha}\left[-\operatorname{grad} p + \varrho^f \boldsymbol{b}\right]. \tag{6-5}$$

If the external body force is conservative, then $\boldsymbol{b}$ is derivable from a potential, i.e.,

$$\boldsymbol{b} = -\operatorname{grad}\Phi\,, \tag{6-6}$$

and thus

$$\boldsymbol{b} = \frac{1}{\alpha}\left[-\operatorname{grad} p - \varrho^f \operatorname{grad}\Phi\right]. \tag{6-7}$$

Now, on making the additional assumption that the fluid is incompressible, i.e., $\varrho^f =$ constant, we obtain

$$\boldsymbol{b} = -\frac{1}{\alpha}\operatorname{grad}\left[p + \varrho^f \Phi\right], \tag{6-8}$$

which is essentially a statement of Darcy's equation.

Bibliography

[1] K.R. Rajagopal and L. Tao, *Mechanics of Mixtures*, World Scientific, Singapore-New Jersey-London-Hong Kong, 1995.

[2] C. Truesdell, *Rational Thermodynamics*, Springer–Verlag, New York, 1984.

[3] R.J. Atkin and R.E. Craine, Q.J.Mech.Appl.Math. **XXIX** (1976), 209.

[4] R.M. Bowen, *Theory of mixtures in Continuum Physics III*, (ed. A.C. Eringen), Academic Press, New York, 1976.

[5] A. Bedford and D.S. Drumheller, Intl.J.Engng.Sci. **21** (1983), 863.

[6] S.L. Passman, J.W. Nunzianto and E.K. Walsch, Rational Thermodynamics by C. Truesdell, Springer–Verlag, New York, 1984.

[7] C. Truesdell, Rend.Lincei. **22** (1957), 33.

[8] C. Truesdell, Rend.Lincei. **22** (1957), 158.

[9] C. Maxwell, *On the dynamical theory of gases*, Phil.Trans.Roy.Soc. **157** (1867), 49.

[10] S. Chapman and T.G. Gowling, *Mathematical theory of Non-uniform Gases*, University Press, Cambridge, 1939.

[11] A. Fick, Ann.der Phys. **94** (1855), 56.

[12] J.G.Kirkwood and F.P. Buff, J.Chem.Phys. **19** (1951), 774.

[13] F.J. Pearson and G.S. Rushbrooke, Proc.Roy.Soc.Edinburgh **A64** (1957), 305.

[14] R.M. Mazo, J.Chem.Phys. **29** (1958), 1112.

[15] R.J.Bearman and J.G. Kirwood, J.Chem.Phys. **28** (1958), 136.

[16] G.A. Mansoori and E. Matteoli, Fluctuation Theory of Mixtures, (eds. G.A. Mansoori and E. Matteoli), Taylor and Francis, New York, 1990.

[17] C. Truesdell and R. Toupin, *The Classical Field Theories in Handbuch der Physik III/1*, (ed. W. Flugge), Springer-Verlag, New York, 1960.

[18] W. Darcy, *Les Fontaines Publiques de La Ville de Dijon*, Victor Dalmont, Paris, 1856.

[19] C. Truesdell, *A first course in Rational Continuum Mechanics*, Academic Press, New York, 1984.

[20] A.E. Green and P.M. Naghdi, Q.J.Mech.Appl.Mech. **XXII** (1967), 427.

[21] I. Samohýl, *Thermodynamics of Irreversible Processes in Fluid Mixtures* (1987), Tuebner-Texte zur Physik, Leipzig.

[22] G. Johnson, M. Massoudi and K.R. Rajagopal, *A review of Interaction Mechanism in Fluid-Solid Flows*, DOE/PETC/TR-90/9, DE91000941, 1991.

[23] N. Mills, Intl.J.Engng.Sci. **4** (1966), 97.

[24] A.E. Green, P.M. Naghdi and J.A. Trapp, Intl.J.Engng.Sci. **8** (1970), 891.

[25] I. Liu, Arch.Ratl.Mech.Anal. **46** (1972), 131.

[26] K. Hutter and B. Svendsen, *Constrained Mixtures*, to appear in Intl.J.Engng.Sci..

[27] J.J. Shi, A.S. Wineman and K.R. Rajagopal, Intl.J.Engng.Sci. **19** (1981), 871.

[28] F.T. Wall and P.J. Flory, J.Chem.Phys. **19** (1951), 1435.

[29] M.V. Gandhi, K.R. Rajagopal and A.S. Wineman, Intl.J.Engng.Sci. **25** (1987), 1441.

[30] F. Dai and K.R. Rajagopal, Acta Mech. **82** (1990), 61.

[31] A. Einstein, Ann. Physik **19** (1906), 286.

[32] A. Einstein, Ann. Physik **34** (1911), 591.

[33] A. Einstein, *Theory of Brownian Movement*, Dover, New York, 1956.

[34] G.B. Jeffrey, Proc.Roy.Soc.London **A102** (1922), 161.

[35] G.I. Taylor, Proc.Roy.Soc.London **A138** (1932), 141.

[36] E. Guth and R. Simha, Kolloid Zeit. **74** (1936), 266.
[37] H. Brenner, Phys.Fluids **1** (1958), 338.
[38] G. Batchelor, J.Fluid Mech. **52** (1972), 245.
[39] J.D. Murray, J.Fluid Mech. **21** (1965), 465.
[40] T.B. Anderson and R. Jackson, Ind.Engng.Chem.Fund. **6** (1967), 527.
[41] M.M.El-Kaissy, *The Thermodynamics of Multiphase Systems with Applications to Fluidized Continua*, Ph. D. thesis, School of Engineering, Stanford University (1975).
[42] J.E. Adkins, Phil.Trans.Roy.Soc.London **A255** (1963), 607.
[43] J.E. Adkins, Phil.Trans.Roy.Soc.London **A255** (1963), 635.
[44] D.A. Drew and L. Segal, Studies in Applied Math. **L** (1971), 205.
[45] K.R. Rajagopal, M. Massoudi and J.M. Ekmann, Recent Developments in Structured Continua II, (eds. D. DeKee and P.N. Kaloni), Longman Scientific and Technical, United Kingdom, 1990.
[46] G. Johnson, M. Massoudi and K.R. Rajagopal, Chem.Engng.Sci. **46** (1991), 1713.
[47] G. Johnson, M. Massoudi and K.R. Rajagopal, Intl.J.Engng.Sci. **29** (1991), 649.
[48] G. Johnson, M. Massoudi and K.R. Rajagopal, Recent Advanmces in Structured Continua, AMD-117, ASME, New York, 1991.
[49] R.E. Craine, Intl.J.Engng.Sci. **9** (1971), 1177.
[50] A.E. Green and P.M. Naghdi, Intl.J.Engng.Sci. **3** (1965), 231.
[51] N. Mills, Q.J.Mech.Appl.Math. **20** (1967), 499.
[52] R. Bowen, Arch.Ratl.Mech.Anal. **24** (1967), 370.
[53] I. Müller, Arch.Ratl.Mech.Anal. **28** (1968), 1.
[54] I. Müller, *Thermodynamics and Statistical Mechanics of Fluids and Mixtures of Fluids*, Lecture Notes of the CNR Scuola Estiva Della Fisica Matematica, Bavi, Cassano, 1976.
[55] W.O. Williams and R. Sampaio, J.Appl.Math.Phys. **28** (1977), 607.
[56] S.L. Soo, *Fluid Dynamics of Multiphase Systems*, Blaisdell, Waltham, 1967.
[57] K. Chamniprasart, A. Al-arabi, K.R. Rajagopal and A.Z. Szeri, J.Tribology **115** (1993), 253.
[58] S.H. Wang, A. Al-arabi, K.R. Rajagopal and A.Z. Szeri, J.Tribology **115** (1993), 515.
[59] A. Al-arabi, S.H. Wang, K.R. Rajagopal and A.Z. Szeri, J.Tribology **115** (1993), 46.

[60] L.P. Eisenhart, *Riemannian Geometry*, University Press, Princeton, 1926.
[61] W. Kosinski, *Field Singularities and Wave Analysis in Continuum Mechanics*, Halstead Press, New York, 1986.

K.R. Rajagopal
Department of Mechanical Engineering
Texas A&M University
College Station
Texas
U.S.A

RALF KAISER and WOLF VON WAHL

A New Functional for the Taylor-Couette Problem in the Small-Gap Limit

0. Introduction, notation

The present paper deals with the Taylor-Couette problem in the small-gap limit, in particular with the stability of the basic flow.

First we discuss in detail the transformation of the equation of motion into a rotating coordinate-system, as well as the small-gap approximation of the basic flow. The resulting system for the perturbation is proved to describe the disturbances of plane Couette-flow in an infinite rotating layer. This connection was already exploited by Busse in [1]. If the interior cylinder has radius R_1 and angular velocity Ω_1, whereas these quantities are R_2, Ω_2 for the exterior one, then the linearized eigenvalue problem at criticality and the marginal eigenvalue problem associated with the energy-functional coincide if

$$(0.1) \qquad \begin{cases} -4\frac{R_2-R_1}{R_1+R_2} = \frac{\Omega_2-\Omega_1}{\Omega_2+\Omega_1} \\ \text{and if the perturbation is axisymmetric,} \end{cases}$$

cf. [1]. As was shown in [9] this is the only possible situation where the marginal cases for monotonic energy-stability and linearized stability concur. As a consequence monotonic energy-stability is followed up by instability if and only if (0.1) is fulfilled. In all other cases the usual gap between monotonic energy-stability and criticality occurs. In particular this is so if the disturbance is axisymmetric and the condition $-4(R_2 - R_1)/(R_1 + R_2) = (\Omega_2 - \Omega_1)/(\Omega_2 + \Omega_1)$ is violated, say by a change of the angular velocities of the rotating cylinders. As was found out by Busse [2], in the small-gap limit there is a close connection between the $2D$-Boussinesq-system for Prandtl-number 1 and the axisymmetric Taylor-Couette-problem. This connection we exploit to construct a new functional which exhibits unconditional stability up to criticality for the axisymmetric Taylor-Couette-problem. The functional in question is a modified energy-functional. For its construction the structure of the nonlinearity has to be used carefully. Conditional stability up to criticality by means of similar functionals has been shown in our situation earlier by Joseph in [5, §42] and by Galdi and Rionero in [3, sec.5.4], cf. also [4, sec.4.A]. When approaching criticality however the stability ball in [3, 4, 5] may become smaller and smaller. Thus subcritical bifurcation is not excluded. The physical meaning of unconditional stability up to criticality is that subcritical bifurcations cannot occur. In our situation this means that if there is any subcritical bifurcation then it has to be necessarily three-dimensional.

When transforming the original problem into the stability problem for plane Couette-flow in an infinite rotating layer the following decomposition of a solenoidal vector field $\mathbf{u}$ can be used to advantage:

$$\mathbf{u} = \text{ curl curl}(\varphi\mathbf{k}) + \text{curl}(\psi\mathbf{k}) + \mathbf{f}.$$

$\mathbf{u}$ is assumed to be periodic in the plane variables x, y, say with wave-numbers α, β in x-direction and y-direction. Thus x, y vary in $[-\frac{\pi}{\alpha}, \frac{\pi}{\alpha}]$, $[-\frac{\pi}{\beta}, \frac{\pi}{\beta}]$ respectively. $\mathbf{k}$ is the unit-vector in z-direction, $z \in [-\frac{1}{2}, \frac{1}{2}]$, and is orthogonal with respect to the layer. φ, ψ are functions having the same periodicity as $\mathbf{u}$. Their mean values over $(-\frac{\pi}{\alpha}, \frac{\pi}{\alpha}) \times (-\frac{\pi}{\beta}, \frac{\pi}{\beta})$ vanish for all z. The vector field $\mathbf{f}$ depends on z only. Its third component f_3 is constant and vanishes if $\mathbf{u}$ vanishes at the walls of the layer $z = \pm\frac{1}{2}$. We use the notations

$$\boldsymbol{\delta}\varphi = \text{ curl curl } \varphi\mathbf{k} = \begin{pmatrix} \partial_{xz}\varphi \\ \partial_{yz}\varphi \\ (-\Delta_2)\varphi \end{pmatrix} \text{ with}$$

$(-\Delta_2) = -\partial_{xx} - \partial_{yy}$ as plane Laplacian,

$$\boldsymbol{\varepsilon}\psi = \text{ curl } \psi\mathbf{k} = \begin{pmatrix} \partial_y\psi \\ -\partial_x\psi \\ 0 \end{pmatrix},$$

$\boldsymbol{\delta} \cdot$ (vector field) $= \partial_{xz}$ (first component of vector field) $+ \ldots,$

and similarly for $\boldsymbol{\varepsilon}\cdot$(vector field). The plane periodicity cell is $\mathcal{P} = (-\frac{\pi}{\alpha}, \frac{\pi}{\alpha}) \times (-\frac{\pi}{\beta}, \frac{\pi}{\beta})$. As layer we address the set $\Omega = \mathcal{P} \times (-\frac{1}{2}, \frac{1}{2})$. If $\mathbf{u}$ is solenoidal and if $\mathbf{u} = 0$ at $z = \pm\frac{1}{2}$ then

(0.2) $$\varphi = \partial_z\varphi = \psi = f_1 = f_2 = 0 \quad \text{at} \quad z = \pm\frac{1}{2}$$

and vice-versa. Steady or unsteady solutions of the Navier-Stokes equations in an infinite rotating layer are now rewritten in terms of the new unknown vector

$$\Phi = \begin{pmatrix} \varphi \\ \psi \\ f_1 \\ f_2 \end{pmatrix}$$

$$\text{if } f_3 = 0, \text{i.e. if } \mathbf{u} = 0 \text{ at } z = \pm\frac{1}{2},$$

$$\Phi = \begin{pmatrix} \varphi \\ \psi \\ f_1 \\ f_2 \\ f_3 \end{pmatrix} \text{ otherwise.}$$

The pressure is thus eliminated and the nonlinearity in the new system is almost local. This new system in the steady case looks like

$$\mathcal{A}\Phi + \mathcal{C}\Phi + \mathcal{M}(\Phi, \Phi) = F. \tag{0.3}$$

$\mathcal{B}, \mathcal{A}$ are differential operators being of higher order when operating on φ, ψ. $\mathcal{C}$ is skew-symmetric and stands for the Coriolis-term. The nonlinearity $\mathcal{M}(.,.)$ is bilinear in Φ as usual. F stands for an external force. If Φ satisfies (0.2) then $\mathcal{B}, \mathcal{A}$ are strictly positive definite selfadjoint operators in $\mathcal{H} = L^2_M(\Omega) \times L^2_M(\Omega) \times L^2((-\frac{1}{2}, \frac{1}{2})) \times L^2((-\frac{1}{2}, \frac{1}{2}))$. The subscript $._M$ indicates mean value 0 over $\mathcal{P}$.

If Φ_s is any steady flow satisfying (0.3) then a disturbance Φ of Φ_s under homogeneous rigid boundary conditions satisfies

$$\left\{ \begin{array}{l} \mathcal{B}\partial_t\Phi + \mathcal{A}\Phi + \mathcal{C}\Phi + \mathcal{M}(\Phi_s, \Phi) + \\ \qquad\qquad + \mathcal{M}(\Phi, \Phi_s) + \mathcal{M}(\Phi, \Phi) = 0, \\ \text{together with boundary-condition (0.2).} \end{array} \right. \tag{0.4}$$

Of course we have to prescribe an initial-value Φ_0 for Φ. The class of solutions of (0.4) we refer to is characterized by

$$\begin{array}{rcl} \Phi' & \in & L^2((0,T), \mathcal{D}(\mathcal{B})), \\ \Phi & \in & L^2((0,T), \mathcal{D}(\mathcal{A})), \end{array}$$

Φ is continuous from $[0, T]$ into a suitable interpolation space between $\mathcal{D}(B)$ and $\mathcal{D}(\mathcal{A})$.

The graph-norm of $\mathcal{A}$ is of course stronger than the graph-norm of $\mathcal{B}$. Thus the initial-value Φ_0, which has to be taken from the interpolation space just mentioned, lies between $\mathcal{D}(\mathcal{A})$ and $\mathcal{D}(\mathcal{B})$. The norm $\|\mathcal{B}^{\frac{1}{2}}\Phi(t)\|^2$ is precisely the kinetic energy of the perturbation at time t and given by

$$\begin{array}{rcl} \|\mathcal{B}^{\frac{1}{2}}\Phi(t)\|^2 & = & \|(-\Delta)^{\frac{1}{2}}(-\Delta_2)^{\frac{1}{2}}\varphi(t)\|^2 + \|(-\Delta_2)^{\frac{1}{2}}\psi(t)\|^2 + \\ & & + \|f_1(t)\|^2 + \|f_2(t)\|^2. \end{array}$$

Norms usually refer to the $\mathcal{H}$-norm or the $L^2(\Omega)$-norm. The sort of solutions of (0.4) we are interested in are in general known to exist only locally in time, this is: $T < T(\Phi_0)$, where $T(\Phi_0)$, $0 < T(\Phi_0) \leq +\infty$, indicates the maximal interval of existence. If Φ does not depend on one of the plane variables x or y, then $T(\Phi_0) = +\infty$. In three dimensions weak solutions of (0.4) are known to exist globally. The positive real axis can be represented as

$$[0, +\infty) = [0, T(\Phi_0)) \cup \bigcup_{\nu=1}^{\infty} I_\nu \cup S$$

where the I_ν are open intervals and S has measure 0. On $[0, T(\Phi_0))$ and on the I_ν the weak solution is a strong one in the sense explained previously. S is called the singular set. It was proved in [6] that if a perturbation is asymptotically stable in the energy-norm as weak solution of (0.4) then one of the intervals I_ν has the form $(a, +\infty)$ and also higher order derivatives of the solution tend to 0 if $t \to +\infty$. This is true for any size of initial values. Observe that kinetic energy is well defined and at least locally (in time) bounded for any weak solution. For the investigation of stability of a steady flow it may thus be sufficient to consider strong solutions of (0.4).

For a detailed explanation of the decomposition $\mathbf{u} = \boldsymbol{\delta}\varphi + \boldsymbol{\varepsilon}\psi + \mathbf{f}$ into a poloidal part $\boldsymbol{\delta}\varphi$, a toroidal one $\boldsymbol{\varepsilon}\psi$ and the mean flow $\mathbf{f}$ and its consequences for the Navier-Stokes equations as well as the Boussinesq equations we refer to [7, sect.0] or [9, sect.0].

I. Plane Couette flow in a rotating coordinate system: physical background

Plane Couette flow in a rotating coordinate system can be realized as Taylor-Couette flow in the small-gap limit. It is the aim of this section to explain this connection in some detail. First, geometry and basic flow of the Taylor-Couette system are presented. Then, the transformation of the basic flow and the equation of motion into a rotating coordinate system is performed. A third paragraph discusses the small-gap approximation. Special emphasis is laid, here, on controlling the approximations made. Finally, the resulting equations are made dimensionless by a suitable scaling.

A) Geometry and basic flow of the Taylor-Couette system

Let us consider an incompressible fluid between two rotating coaxial cylinders with radii R_1 and R_2, respectively (see Fig.1). The fluid sticks to the cylinder walls which rotate with angular velocities Ω_1 and Ω_2, respectively. For later use we define a mean radius R and gap width d,

$$R := \frac{1}{2}(R_1 + R_2), \quad d := (R_2 - R_1),$$

as well as a mean angular velocity Ω and a relative angular velocity $\Delta\Omega$,

$$\Omega := \frac{1}{2}(\Omega_1 + \Omega_2), \quad \Delta\Omega := \Omega_2 - \Omega_1.$$

The Navier-Stokes equations take in a fixed frame S for an incompressible fluid the form:

$$\text{(I.1)} \qquad \partial_t \mathbf{v} - \nu\Delta\mathbf{v} + \mathbf{v}\cdot\nabla\mathbf{v} + \nabla p = \mathbf{F}, \qquad \nabla\cdot\mathbf{v} = 0.$$

Here, $\mathbf{v}$ denotes the velocity field, ν the viscosity, p the pressure and $\mathbf{F}$ possible external forces, such as gravity. Due to incompressibility the fluid density ϱ is constant; for simplicity, it is set to one, $\varrho \equiv 1$. For convenience polar coordinates s, ϕ are introduced

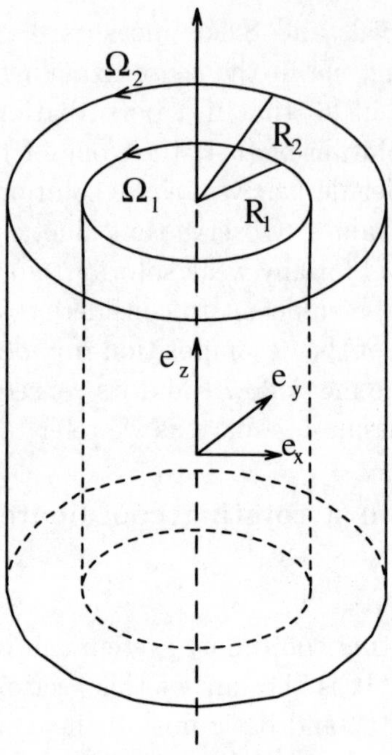

Fig. 1

in the x, y-plane. Together with z they form a cylindrical coordinate system, in which e.g. the gradient has the following form:

$$\nabla = \mathbf{e}_s\partial_s + \mathbf{e}_\phi\frac{1}{s}\partial_\phi + \mathbf{e}_z\partial_z.$$

Note that the basis vectors $\mathbf{e}_s, \mathbf{e}_\phi$ are not constant,

(I.2) $$\partial_\phi\mathbf{e}_s = \mathbf{e}_\phi, \quad \partial_\phi\mathbf{e}_\phi = -\mathbf{e}_s.$$

The basic flow is driven by the moving cylinder walls; consequently, it should have only a ϕ-component. Furthermore, the cylinders are assumed to be sufficiently long so that perturbations from top or bottom can be neglected and a solution homogeneous in the z-direction can be expected. We are seeking thus a steady solution $\mathbf{v}_0, p_0$ of Eq.(I.1) with $\mathbf{F} \equiv 0$, where $\mathbf{v}_0$ is of the form $\mathbf{v}_0 = v_0(s,\phi)\mathbf{e}_\phi$ and satisfies the two boundary conditions

(I.3) $$v_0|_{s=R_i} = \Omega_i R_i, \quad i = 1, 2.$$

From Eq.(I.1b) we have immediately $\partial_\phi v_0 = 0$, i.e. v_0 is a function of s alone. Using Eqs.(I.2) the nonlinear and the dissipative term in Eq.(I.1a) furnish

$$\begin{aligned}\mathbf{v}_0 \cdot \nabla\mathbf{v}_0 &= -\frac{1}{s}v_0^2\,\mathbf{e}_s,\\ \Delta\mathbf{v}_0 &= \partial_s[\frac{1}{s}\partial_s(sv_0)]\mathbf{e}_\phi,\end{aligned}$$

so that Eq.(I.1a) takes the form:

$$\{-\frac{1}{s}v_0^2 + \partial_s p\}\mathbf{e}_s + \{-\nu\partial_s[\frac{1}{s}\partial_s(sv_0)] + \frac{1}{s}\partial_\phi p\}\mathbf{e}_\phi + \partial_z p\,\mathbf{e}_z = 0.$$

From the z-component we have that p is independent of z, integrating the ϕ-component with respect to ϕ and using periodicity of p in ϕ furnish

$$p_0 = p_0(s),$$

(I.4) $$\partial_s[\frac{1}{s}\partial_s(sv_0)] = 0,$$

and integrating the s-component leads, finally, to

(I.5) $$p_0 = \text{ const } + \int_{R_1}^{s} \frac{1}{\sigma} v_0^2(\sigma)\, d\sigma.$$

The general solution of (I.4) is

(I.6) $$v_0(s) = A\,s + \frac{B}{s}.$$

The constants A and B are determined by the boundary conditions (I.3),

(I.7)
$$A = \frac{R_2^2\Omega_2 - R_1^2\Omega_1}{R_2^2 - R_1^2} = \frac{R_2^2\Omega_2 - R_1^2\Omega_1}{2Rd},$$
$$B = -\frac{R_1^2R_2^2(\Omega_2 - \Omega_1)}{R_2^2 - R_1^2} = -\frac{R_1^2R_2^2\Delta\Omega}{2Rd}.$$

Eqs.(I.5,6,7) constitute the basic flow we have looked for.

Remark: From a physical point of view the pressure gradient $\partial_s p_0$ furnishes just the centripetal force $\frac{v_0^2}{s}$ which is needed to constrain the fluid on circular orbits.

B) Transformation in a rotating coordinate system

Let $S'(\mathbf{e}'_x, \mathbf{e}'_y, \mathbf{e}'_z)$ be a new coordinate system which rotates with constant angular velocity Ω with respect to $S(\mathbf{e}_x, \mathbf{e}_y, \mathbf{e}_z)$ about the joint axis $\mathbf{e}_z = \mathbf{e}'_z$ (see Fig.2). Old and new coordinates are related by

(I.8) $$s' = s, \quad \phi' = \phi - \Omega t, \quad z' = z.$$

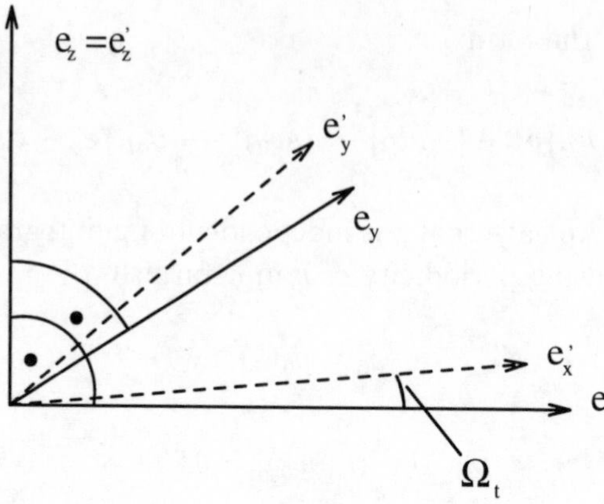

Fig. 2

Proposition I.1

$$\text{(I.9)} \qquad \begin{aligned} \mathbf{v}'(s',\phi',z',t) &:= \mathbf{v} - \mathbf{v}_{Rel}, \\ p'(s',\phi',z',t) &:= p - \tfrac{1}{2}|\mathbf{v}_{Rel}|^2, \end{aligned}$$

with $\mathbf{v}_{Rel} := \mathbf{\Omega} \times \mathbf{r}'$, $\mathbf{\Omega} = \Omega \mathbf{e}_z$ *are a solution of the equation*

$$\text{(I.10)} \qquad \partial_t \mathbf{v}' - \nu \Delta' \mathbf{v}' + \mathbf{v}' \cdot \nabla' \mathbf{v}' + 2\,\mathbf{\Omega} \times \mathbf{v}' + \nabla' p' = \mathbf{F}, \quad \nabla' \cdot \mathbf{v}' = 0,$$

in the rotating coordinate system S' *if and only if* $\mathbf{v}, p$ *are a solution of Eq.*(I.1) *in* S.

Note that on the right-hand side of Eqs.(I.9) as well as in $\mathbf{F}$ the old variables have been expressed by the new ones via (I.8).

Proof: Eqs.(I.1a) and (I.10a) are equivalent if the following identities are taken into account:

$$\begin{aligned} \partial_t \mathbf{v}' &= \partial_t \mathbf{v} + \Omega(\partial'_\phi v_\mu)\mathbf{e}'_\mu, \\ \Delta' \mathbf{v}_{Rel} &= 0, \\ \mathbf{v} \cdot \nabla' \mathbf{v}_{Rel} &= \mathbf{v} \cdot \nabla'(\mathbf{\Omega} \times \mathbf{r}') = v_k \partial'_k \varepsilon_{lmn} \Omega_l r'_m \mathbf{e}'_n = \varepsilon_{lmn} \Omega_l v_m \mathbf{e}'_n = \mathbf{\Omega} \times \mathbf{v}, \\ \mathbf{v}_{Rel} \cdot \nabla' \mathbf{v} &= \Omega \partial'_\phi (v_\mu \mathbf{e}'_\mu) = \Omega(\partial'_\phi v_\mu)\mathbf{e}'_\mu + \mathbf{\Omega} \times \mathbf{v}, \\ \mathbf{v}_{Rel} \cdot \nabla' \mathbf{v}_{Rel} &= -\Omega^2 s\, \mathbf{e}'_s, \\ \mathbf{\Omega} \times \mathbf{v}_{Rel} &= -\Omega^2 s\, \mathbf{e}'_s. \end{aligned}$$

Here, $\mathbf{v}_{Rel}$ in cylindrical coordinates, $\mathbf{v}_{Rel} = \Omega s \mathbf{e}'_\phi$, has been used. Latin indices $k, l, ...$ refer to Cartesian coordinates and Greek indices $\mu, \nu, ...$ to cylindrical ones. Summation over double indices is assumed. Observe that ∂_t on the right-hand side of the first identity refers only to the explicit time dependence in $\mathbf{v}(\mathbf{r}', t)$.

Eqs.(I.1b) and (I.10b) are trivially equivalent. □

The basic flow $\mathbf{v}_0$ does neither depend on t nor ϕ, i.e. a distinction between old and new variables is unnecessary. With the definition

$$\text{(I.11)} \qquad \begin{aligned} \mathbf{v}_1 &:= \mathbf{v}_0 - \mathbf{v}_{Rel}, \\ p_1 &:= p_0 - \tfrac{1}{2}|\mathbf{v}_{Rel}|^2 \end{aligned}$$

$\mathbf{v}_1, p_1$ satisfy the steady equation (I.10) with $\mathbf{F} \equiv 0$.

For later use we note yet the equations for a perturbation $\delta\mathbf{v} = \mathbf{v} - \mathbf{v}_1$, $\delta p = p - p_1$ of the basic solution $\mathbf{v}_1, p_1$ in S':

$$\text{(I.12)} \qquad \begin{aligned} \partial_t \delta\mathbf{v} - \nu\Delta'\delta\mathbf{v} + \mathbf{v}_1 \cdot \nabla'\delta\mathbf{v} + \delta\mathbf{v} \cdot \nabla'\mathbf{v}_1 + \delta\mathbf{v} \cdot \nabla'\delta\mathbf{v} & \\ +2\,\mathbf{\Omega} \times \delta\mathbf{v} + \nabla'\delta p &= 0, \\ \nabla' \cdot \delta\mathbf{v} &= 0. \end{aligned}$$

Remarks: 1) From the proof of Proposition I.1 one sees that a steady flow in S is in general no longer steady in S'. Only in case of axisymmetry (and this case applies to the basic flow!) steadiness in S implies steadiness in S'.

2) The pressure p_1 measured in S' is reduced by the centrifugal term $\frac{1}{2}\Omega^2 s^2$ compared to the pressure p_0 measured in S. This means the centrifugal term is absorbed in the pressure term and the only term in Eq.(I.10) which relates to the rotating coordinate system is the Coriolis term $2\,\mathbf{\Omega} \times \mathbf{v}'$.

C) Approximation of the basic flow in the small-gap limit

Two approximations are performed and their validity is discussed. The first one breaks off the Taylor expansion of the term $\frac{B}{s}$ in v_0 about the gap center R after first order:

$$\frac{B}{s} = B[\frac{1}{R} - \frac{s-R}{R^2} + T_1(s)].$$

Here, $T_1(s)$ denotes Taylor's remainder

$$T_1(s) = \int_R^s \left(\frac{1}{\sigma}\right)''(s-\sigma)\,d\sigma = 2\int_R^s \frac{s-\sigma}{\sigma^3}\,d\sigma = \frac{(s-R)^2}{sR^2}.$$

The approximation is justified if T_1 is small compared with the first order term,

$$\frac{(s-R)^2}{sR^2} << \frac{|s-R|}{R^2},$$

i.e.

$$\text{(I.13)} \qquad \left|\frac{s-R}{s}\right| << 1.$$

With $|\frac{s-R}{s}| \le \frac{d}{2R_1}$ and $R \le 2R_1$ Eq.(I.13) is implied by the small-gap condition

(I.14)
$$\frac{d}{R} << 1.$$

The second approximation replaces the mean radius R by the geometric mean $\sqrt{R_1 R_1}$ in the denominators of $\tilde{v}_0(s) = A\,s - \frac{B}{R^2}s + 2\frac{BR^3}{R^4}$, i.e.

(I.15)
$$\tilde{\tilde{v}}_0(s) = A\,s - \frac{B}{R_1 R_2}s + 2\frac{BR^3}{R_1^2 R_2^2}.$$

The approximation is justified if

(I.16)
$$\left|B\frac{s}{R^2} - B\frac{s}{R_1 R_2}\right| << \left|A\,s - \frac{B}{R_1 R_2}s\right|$$

and

(I.17)
$$\left|\frac{R^3}{R_1^2 R_2^2} - \frac{1}{R}\right| << \frac{R^3}{R_1^2 R_2^2}.$$

(I.17) is equivalent to

$$\left|\frac{R^4 - R_1^2 R_2^2}{R^4}\right| = \left|\frac{(R^2 - R_1 R_2)(R^2 + R_1 R_2)}{R^4}\right| << 1.$$

With $R^2 = R_1 R_2 + \frac{d^2}{4} \ge R_1 R_2$ (I.17) is thus satisfied if

$$\frac{1}{2}\frac{d^2}{R^2} << 1,$$

which is implied by the small-gap condition (I.14). (I.16) is equivalent to

$$|B|\frac{d^4}{4R^2 R_1 R_2} << \left|A - \frac{B}{R_1 R_2}\right|.$$

With A and B from Eqs.(I.7) we have

$$R_1 R_2 \frac{d^2}{4R^2}|\Delta\Omega| << |R_2^2\Omega_2 - R_1^2\Omega_1 + R_1 R_2 \Delta\Omega|,$$

and using the identity

(I.18)
$$R_2^2\Omega_2 - R_1^2\Omega_1 = (2R^2 - R_1 R_2)\Delta\Omega + 2Rd\Omega,$$

(I.16) is satisfied if

(I.19)
$$d^2|\Delta\Omega| << 2R|R\Delta\Omega + d\Omega|.$$

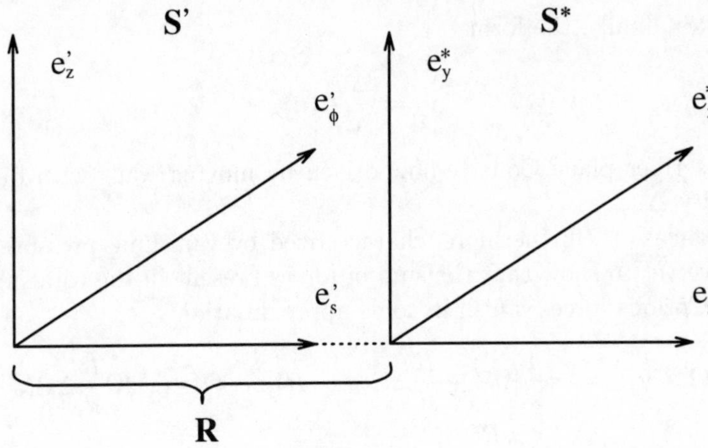

Fig. 3

For $\Omega \cdot \Delta\Omega \geq 0$ (I.19) is again implied by the small-gap condition (I.14), for $\Omega \cdot \Delta\Omega < 0$, however, the additional condition

$$\frac{d^2}{R^2} << \left|1 + \frac{d}{R}\frac{\Omega}{\Delta\Omega}\right| \tag{I.20}$$

has to be considered.

In summary, if conditions (I.14) and (I.20) are satisfied the approximation of v_0 by $\tilde{\tilde{v}}_0$ is justified. Using again Eqs.(I.7) and (I.18) one obtains for $\tilde{\tilde{v}}_0$:

$$\begin{aligned} \tilde{\tilde{v}}_0(s) &= \frac{1}{2Rd}[(R_2^2\Omega_2 - R_1^2\Omega_1 + R_1R_2\Delta\Omega)s - 2R^3\Delta\Omega] \\ &= \Omega s + \frac{R}{d}\Delta\Omega(s - R), \end{aligned}$$

and from Eq.(I.11a) for $\tilde{\tilde{\mathbf{v}}}_1$:

$$\tilde{\tilde{\mathbf{v}}}_1 = \frac{R}{d}\Delta\Omega(s' - R)\mathbf{e}'_\phi. \tag{I.21}$$

In the small-gap limit the cylindrical coordinate system S' can locally be approximated by a plane one S^*. If we use the identification (see Fig. 3)

$$\begin{aligned} \mathbf{e}'_\phi &\to \mathbf{e}^*_x, \\ \mathbf{e}'_s &\to \mathbf{e}^*_z \text{ (shifted by } R), \\ \mathbf{e}'_z &\to \mathbf{e}^*_y, \end{aligned} \tag{I.22}$$

and if we observe that the relative velocity Δv of the cylinder walls in the rotating coordinate system S' is

$$\Delta v = R_2(\Omega_2 - \Omega) - R_1(\Omega_1 - \Omega) = R\Delta\Omega,$$

Eq. (I.21) takes, finally, the form

$$\tilde{\tilde{\mathbf{v}}}_1 = \frac{\Delta v}{d} z^* \mathbf{e}_x^*. \tag{I.23}$$

Eq. (I.23) describes plane Couette-flow driven by moving walls with distance d and relative velocity Δv.

The Couette-case is, furthermore, characterized by vanishing pressure. To be consistent we have yet to show that the sum of forces present in the rotating system, i.e. pressure and Coriolis force, vanish in some approximation:

$$\begin{aligned} \nabla' \tilde{\tilde{p}}_1 + 2\,\boldsymbol{\Omega} \times \tilde{\tilde{\mathbf{v}}}_1 &= \{\frac{1}{s'}[\Omega s' + \frac{R}{d}\Delta\Omega(s'-R)]^2 - \Omega^2 s' - 2\Omega\frac{R}{d}\Delta\Omega(s'-R)\}\mathbf{e}'_s \\ &= \Delta\Omega^2 \frac{R^2}{d^2}\frac{{z^*}^2}{R+z^*}. \end{aligned} \tag{I.24}$$

This expression is $O({z^*}^2)$, which is consistent with the linear approximation (in z^* around 0) performed on the basic flow. If we require

$$|\Delta\Omega| << |\Omega| \tag{I.25}$$

expression (I.24) is in fact much smaller than the individual forces present in the rotating system, i.e. Coriolis - and centrifugal force and $\nabla' \tilde{\tilde{p}}_0$:

$$|\nabla' \tilde{\tilde{p}}_1 + 2\,\boldsymbol{\Omega} \times \tilde{\tilde{\mathbf{v}}}_1| << \begin{cases} |2\,\boldsymbol{\Omega} \times \tilde{\tilde{\mathbf{v}}}_1| \\ |\Omega^2(R+z^*)| \\ |\nabla' \tilde{\tilde{p}}_0|. \end{cases}$$

Remarks: 1) The only physical situation in which condition (I.20) is relevant is the case of both cylinders moving in the same direction with the inner one moving slightly faster. In this situation the approximation is bad if the system is close to the "resonance case" $d\Omega + R\Delta\Omega = 0$. Observe that the case of coincidence of monotonic energy stability and criticality differs from the resonance case by a factor of 4.

2) The approximation of the cylindrical coordinate system S' by a Cartesian one S^* is, of course, only locally, i.e. for $|\phi'| << 1$, valid. Stability considerations are, in fact, local. A characteristic length in the ϕ'-direction is given by the periodicity length l_ϕ of a periodic perturbation. If l_ϕ can be approximately estimated by d, the angle ϕ necessary to describe a periodic perturbation can be estimated by

$$|\phi| \leq 2\frac{l_\phi}{R} \leq 2\frac{d}{R} << 1,$$

which is implied by the small-gap condition (I.14). A sufficient condition for (I.22) to be valid is thus

$$l_\phi \tilde{<} d, \tag{I.26}$$

i.e. the periodicity length is not much greater than the gap width. Note that in general condition (I.26) can be verified only a-posteriori; for axisymmetric perturbations, however, condition (I.26) is trivially satisfied.

D) Scaling of the Navier-Stokes equations

In order to introduce dimensionless variables in the Navier-Stokes equations we choose the gap width d as length scale and the relative velocity $|\Delta v|$ of the walls as velocity scale. It is appropriate for laboratory experiments to take the viscous scale $\frac{d^2}{\nu}$ as time scale. All other quantities are appropriately scaled with d and Δv.

In Eq.(I.10a) the mass density ϱ is now restored; in the coordinate system S^* the equation has the same form as in S' (stars are henceforth omitted):

$$\partial_t \mathbf{v} - \nu \Delta \mathbf{v} + \mathbf{v} \cdot \nabla \mathbf{v} + 2\,\mathbf{\Omega} \times \mathbf{v} + \nabla(\frac{p}{\varrho}) = \frac{\mathbf{F}}{\varrho}. \tag{I.27}$$

With the scaled quantities $\bar{t} = \frac{\nu}{d^2}t$, $\bar{\mathbf{r}} = \frac{\mathbf{r}}{d}$, $\bar{\mathbf{v}} = \frac{\mathbf{v}}{|\Delta v|}$, $\bar{\mathbf{\Omega}} = \frac{d}{|\Delta v|}\mathbf{\Omega}$, $\overline{(\frac{p}{\varrho})} = \frac{1}{|\Delta v|^2}(\frac{p}{\varrho})$ and $\overline{(\frac{\mathbf{F}}{\varrho})} = \frac{d}{|\Delta v|^2}(\frac{\mathbf{F}}{\varrho})$ Eq.(I.27) transforms into

$$\partial_{\bar{t}}\bar{\mathbf{v}} - \bar{\Delta}\bar{\mathbf{v}} + Re[\bar{\mathbf{v}} \cdot \bar{\nabla}\bar{\mathbf{v}} + 2\bar{\mathbf{\Omega}} \times \bar{\mathbf{v}} + \bar{\nabla}\overline{(\frac{p}{\varrho})}] = Re\,\overline{\left(\frac{\mathbf{F}}{\varrho}\right)},$$

where Re denotes the dimensionless quantity

$$Re := \frac{d|\Delta v|}{\nu}. \tag{I.28}$$

The basic flow (I.23) reads now

$$\bar{\mathbf{v}}_1 = \ \mathrm{sign}(\Delta v)\bar{z}\mathbf{e}_x, \tag{I.29}$$

and the equation for a perturbation $\delta\bar{\mathbf{v}} = \bar{\mathbf{v}} - \bar{\mathbf{v}}_1$ reads

$$\begin{aligned} \partial_{\bar{t}}\delta\bar{\mathbf{v}} - \bar{\Delta}\delta\bar{\mathbf{v}} + Re[\delta\bar{\mathbf{v}} \cdot \bar{\nabla}\bar{\mathbf{v}}_1 + \bar{\mathbf{v}}_1 \cdot \bar{\nabla}\delta\bar{\mathbf{v}} + \delta\bar{\mathbf{v}} \cdot \bar{\nabla}\delta\bar{\mathbf{v}} & \\ +2\,\bar{\mathbf{\Omega}} \times \delta\bar{\mathbf{v}} + \bar{\nabla}\delta\overline{(\frac{p}{\varrho})}] &= 0. \end{aligned} \tag{I.30}$$

With the further scaling $\delta\bar{\bar{\mathbf{v}}} = Re\,\delta\bar{\mathbf{v}}$, $\bar{\bar{\mathbf{\Omega}}} = Re\,\bar{\mathbf{\Omega}}$ and $\delta\overline{\overline{(\frac{p}{\varrho})}} = Re^2\,\delta\overline{(\frac{p}{\varrho})}$ Eq.(I.30) takes the form

$$\begin{aligned} \partial_{\bar{t}}\delta\bar{\bar{\mathbf{v}}} - \bar{\Delta}\delta\bar{\bar{\mathbf{v}}} + Re[\delta\bar{\bar{\mathbf{v}}} \cdot \bar{\nabla}\bar{\mathbf{v}}_1 + \bar{\mathbf{v}}_1 \cdot \bar{\nabla}\delta\bar{\bar{\mathbf{v}}}] + \delta\bar{\bar{\mathbf{v}}} \cdot \bar{\nabla}\delta\bar{\bar{\mathbf{v}}} & \\ +2\,\bar{\bar{\mathbf{\Omega}}} \times \delta\bar{\bar{\mathbf{v}}} + \bar{\nabla}\delta\overline{\overline{(\frac{p}{\varrho})}} &= 0. \end{aligned} \tag{I.31}$$

For later convenience $\text{sign}(\Delta v) = -1$ is chosen. If Eq. (I.29) is inserted in Eq.(I.31), the perturbation equation takes, finally, the form (bars are now omitted):

$$\text{(I.32)} \qquad \begin{aligned} &\partial_t \delta\mathbf{v} - \Delta\delta\mathbf{v} + Re[-\delta v_z \mathbf{e}_x - z\partial_x \delta\mathbf{v}] + \delta\mathbf{v}\cdot\nabla\delta\mathbf{v} \\ &\qquad + 2\boldsymbol{\Omega}\times\delta\mathbf{v} + \nabla\delta(\frac{p}{\varrho}) \;=\; 0, \end{aligned}$$

and the parameters Re and Ω still present in Eq.(I.32) have, if expressed in terms of Ω_1, Ω_2, R and d, now the form:

$$\text{(I.33)} \qquad Re = \frac{Rd}{\nu}|\Omega_2 - \Omega_1|, \qquad \Omega = \frac{d^2}{2\nu}(\Omega_1 + \Omega_2).$$

Remark: There is some freedom in the choice of a scaling. More precisely, scaling of any quantity in Eq.(I.32) with an arbitrary function of Re is still admissible. For example, time is often scaled with the convection scale $\frac{d}{|\Delta v|}$. The ratio of viscous and convection scale is just Re.

II. Unconditional stability up to criticality

In the present section we want to prove that in the small-gap limit the basic flow (in its approximated form) is unconditionally stable against axisymmetric perturbations as long as we stay below criticality. This is exhibited by a suitable functional which is a modified energy functional. At criticality itself this new functional is monotonically non-increasing. For its construction we have to make strong use of the structure of the nonlinearity. The latter is exploited most effectively when we use the decomposition $\mathbf{u} = P + T + F$ addressed to in the introduction. Instead of $\delta\mathbf{v}$ in (I.32) we write $\mathbf{u}$. Set $\boldsymbol{\Omega} = \Omega\mathbf{j}$. Then (I.32) takes the form (cf. [9, pp.119,120]).

$$\text{(II.1)} \qquad \begin{cases} (-\Delta)(-\Delta_2)\partial_t\varphi + \Delta^2(-\Delta_2)\varphi - 2(-\Delta_2)\,\Omega\,\partial_y\psi + \\ \qquad + \boldsymbol{\delta}\cdot(\mathbf{u}\cdot\nabla\mathbf{u}) = 0, \\ (-\Delta_2)\partial_t\psi + (-\Delta)(-\Delta_2)\psi - Re(-\partial_y)(-\Delta_2)\varphi + \\ \qquad + 2\Omega\,(-\partial_y)(-\Delta_2)\varphi - \\ \qquad - \boldsymbol{\varepsilon}\cdot(\mathbf{u}\cdot\nabla\mathbf{u}) = 0, \\ \partial_t f_1 + (-\partial_z^2)f_1 + \frac{1}{|\mathcal{P}|}\int_{\mathcal{P}}(\tilde{\mathbf{u}}\cdot\nabla\tilde{\mathbf{u}})_x\,dx\,dy = 0, \\ \partial_t f_2 + (-\partial_z^2)f_2 + \frac{1}{|\mathcal{P}|}\int_{\mathcal{P}}(\tilde{\mathbf{u}}\cdot\nabla\tilde{\mathbf{u}})_y\,dx\,dy = 0 \end{cases}$$

in the axisymmetric case, this is: $\partial_x\mathbf{u} \equiv 0$ or $\partial_x\varphi = \partial_x\psi \equiv 0$. We have

$$\text{(II.2)} \qquad \mathbf{u} = \boldsymbol{\delta}\varphi + \boldsymbol{\varepsilon}\psi + \mathbf{f}$$

with $\mathbf{f} = (f_1, f_2, 0)^T$,

$$\text{(II.3)} \qquad \tilde{\mathbf{u}} = \boldsymbol{\delta}\varphi + \boldsymbol{\varepsilon}\psi.$$

If β is the wave number in y-direction then $\mathcal{P} = (-\frac{\pi}{\beta}, \frac{\pi}{\beta})$, $|\mathcal{P}| = 2\pi/\beta$. The boundary conditions are rigid ones throughout, this is $\mathbf{u} = 0$ at $z = \pm\frac{1}{2}$. In terms of φ, ψ, f_1, f_2 we obtain

$$(\text{II}.4) \qquad \begin{cases} \varphi = \partial_z\varphi = 0 \text{ at } z = \pm\frac{1}{2}, \\ \psi = 0 \text{ at } z = \pm\frac{1}{2}, \\ f_1 = f_2 = 0 \text{ at } z = \pm\frac{1}{2}. \end{cases}$$

The nonlinear problem (II.1-4) for

$$(\text{II}.5) \qquad \Phi = (\varphi, \psi, f_1, f_2)^T$$

in the axisymmetric case has a unique global (in time) strong solution for any initial value Φ_0 (cf. [10, Chap.IV] and section 0 here).

By $R_{\min}(\beta^2)$, $\beta > 0$, we mean the onset curve of the rigid x-independent Bénard-problem. It can be easily extracted from the onset surface of the rigid Bénard-problem as given in [8]. This curve was already used in [9, pp.122,123] in connection with the stability-problem for plane Couette-flow in an infinite rotating layer. In view of the particular form (I.32) we have given to the Taylor-Couette problem it is not surprising that $R_{\min}(\beta^2)$ is needed here again.

In the first step we determine the critical Reynolds-number for axisymmetric perturbations. We have

Theorem II.1 *After application of $(-\Delta_2)^{-1}$ the linearized eigenvalue-problem for axisymmetric perturbations reads*

$$(\text{II}.6) \qquad \begin{cases} \sigma(-\Delta)\varphi &= \Delta^2\varphi - 2\Omega\,\partial_y\psi, \\ \sigma\psi &= (-\Delta)\psi - (2\Omega - Re)\partial_y\varphi, \\ \sigma f_1 &= (-\partial_z^2) f_1, \\ \sigma f_2 &= (-\partial_z^2) f_2. \end{cases}$$

If

$$2\Omega(Re - 2\Omega) < R_{\min}(\beta^2)$$

then $\xi_0 = \min\{\mathcal{R}e\,\sigma \,|\, \sigma$ *eigenvalue in* (II.6)$\}$ *is positive. If*

$$2\Omega(Re - 2\Omega) > R_{\min}(\beta^2)$$

then $\xi_0 < 0$. *Thus criticality* $\xi_0 = 0$ *occurs if and only if*

$$(\text{II}.7) \qquad 2\Omega(Re - 2\Omega) = R_{\min}(\beta^2).$$

We have an exchange of stability at criticality since there are no purely imaginary eigenvalues. The number

$$Re_c = 2\Omega + \frac{R_{\min}(\beta^2)}{2\Omega},$$

inferred from (II.7), *is called the critical Reynolds-number for axisymmetric perturbations* ($\Omega \neq 0$).

Proof: Set

$$\begin{aligned} R &= 2\Omega(Re - 2\Omega), \\ \hat{\psi} &= 2\Omega\,\partial_y\psi. \end{aligned}$$

There arises

$$\text{(II.8)} \qquad \begin{cases} \sigma(-\Delta)\varphi &= \Delta^2\varphi - \hat{\psi}, \\ \sigma\hat{\psi} &= (-\Delta)\hat{\psi} - R(-\partial_y^2\varphi), \\ \sigma f_1 &= (-\partial_z^2)f_1, \\ \sigma f_2 &= (-\partial_z^2)f_2. \end{cases}$$

If $R > 0$ we set $\vartheta = (\sqrt{R})^{-1}\hat{\psi}$ and arrive at the linearized Bénard-problem

$$\text{(II.9)} \qquad \begin{cases} \sigma(-\Delta)(-\Delta_2)\varphi &= \Delta^2(-\Delta_2)\varphi - \sqrt{R}(-\Delta_2)\vartheta, \\ \sigma\vartheta &= (-\Delta)\vartheta - \sqrt{R}(-\Delta_2)\varphi, \\ \sigma f_1 &= (-\partial_z^2)f_1, \\ \sigma f_2 &= (-\partial_z^2)f_2. \end{cases}$$

for x-independent eigenvectors and for $\Pr = 1$. Thus σ is real. If $R < 0$ we set $\vartheta = \left(\sqrt{|R|}\right)^{-1}\hat{\psi}$.

$$\text{(II.10)} \qquad \begin{cases} \sigma(-\Delta)(-\Delta_2)\varphi &= \Delta^2(-\Delta_2)\varphi - \sqrt{|R|}(-\Delta_2)\vartheta, \\ \sigma\vartheta &= (-\Delta)\vartheta + \sqrt{|R|}(-\Delta_2)\varphi, \\ \sigma f_1 &= (-\partial_z^2)f_1, \\ \sigma f_2 &= (-\partial_z^2)f_2. \end{cases}$$

In that case σ is not necessarily real but $\Re\,\sigma$ is positive. If $R = 0$ we obtain $\sigma\hat{\psi} = (-\Delta)\hat{\psi}$ for an eigenvector $(\varphi, \hat{\psi}, f_1, f_2)^T$ in (II.8). If $\hat{\psi} \neq 0$ it follows that σ is real and positive. If $\hat{\psi} = 0$ then at least one of the quantities φ, f_1, f_2 is $\neq 0$ and we arrive again at the conclusion that $\sigma > 0$. As for the linearized Bénard-problem (II.9) we refer to [9, p.123] where $\frac{1}{2}Re$ should be replaced by $\sqrt{R}$. We take the opportunity to mention one correction in [9, p.123], which is connected with the present argument: In [9, (2.14)] it should read $\sigma(-\partial_y^2\varphi)$ instead of $\sigma\varphi$. □

As it is known from [9, chap.2] the only possibility in (I.32) that monotonic energy stability is followed up by instability is

$$\text{(II.11)} \qquad \begin{cases} Re &= 4\Omega, \\ \partial_x\mathbf{u} &\equiv 0. \end{cases}$$

Observe that this is the only possibility amongst all $3D$-perturbations. As for the Taylor-Couette-problem in the small-gap limit the requirement (II.11) means

$$\text{(II.12)} \qquad \begin{cases} -4\frac{R_2-R_1}{R_1+R_2} = \frac{\Omega_2-\Omega_1}{\Omega_2+\Omega_1}, \\ \mathbf{u} \text{ is axisymmetric.} \end{cases}$$

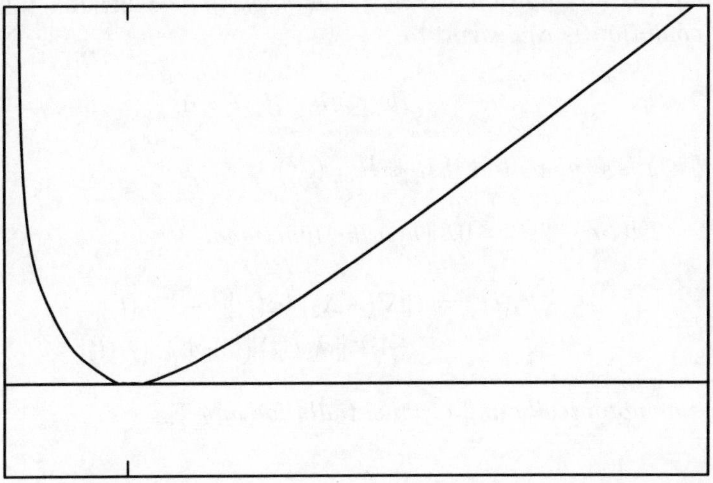

Fig. 4

If for instance Ω_1, Ω_2 are subject to slight changes the first condition in (II.12) is violated. Then necessarily also in the axisymmetric case there arises the usual gap between monotonic energy-stability and linearized stability (cf. Theorem 2.1 in [9]). The energetic Reynolds number Re_E is not influenced by the Coriolis-term. Its value is given by

$$Re_E = 2\sqrt{R_{\min}(\beta^2)}.$$

The case of interest in Theorem II.1 is $\Omega > 0$. Then Re_c assumes its minimal value $2\sqrt{R_{\min}(\beta^2)}$ if and only if $\Omega = \frac{1}{2}\sqrt{R_{\min}(\beta^2)}$. In that case, and only in that,

$$\begin{aligned} Re_E &= Re_c = 4\Omega, \\ &= 2\sqrt{R_{\min}(\beta^2)}. \end{aligned}$$

Fig. 4 explains the situation.

Although thus $Re_c > Re_E$ unless we have $Re = 4\Omega$, there holds

Theorem II.2 *Let $\Phi = (\varphi, \psi, f_1, f_2)^T$ be the unique global (in time) strong solution of* (II.1,2,3,4) *with initial value Φ_0.*

i) Let $R = 2\Omega(Re - 2\Omega) \neq 0$. Then the functional

$$\begin{aligned} F_1(t) &= \|\nabla(-\Delta_2)^{\frac{1}{2}}\varphi(t)\|^2 + \|f_2(t)\|^2 + \\ &\quad + \frac{4\Omega^2}{|2\Omega(Re - 2\Omega)|}(\|\partial_y\psi(t)\|^2 + \|f_1(t)\|^2) \end{aligned}$$

is monotonically non-increasing for any Φ_0 *if* $R = 2\Omega(Re - 2\Omega) \leq R_{\min}(\beta^2)$. *This condition is equivalent to*

$$Re \leq Re_c \ \text{ if } \Omega > 0.$$

The decay is exponential if $R < R_{\min}(\beta^2)$.

ii) Let $R = 2\Omega(Re - 2\Omega) = 0$. *Then the functional*

$$\begin{aligned} F_2(t) &= \|\nabla(-\Delta_2)^{\frac{1}{2}}\varphi(t)\|^2 + \|f_2(t)\|^2 + \\ &\quad + 4\Omega^2\|\partial_y\psi(t)\|^2 + 4\Omega^2\|f_1(t)\|^2 \end{aligned}$$

decays monotonically and exponentially for any Φ_0.

Proof: The second row in (II.1) is multiplied with $2\Omega\gamma$. Then we take the scalar product with $2\Omega\gamma\psi$ over the layer. $\gamma \neq 0$ is a constant to be determined later on. Observe that $\Delta_2 = \partial_y^2$. We arrive at

$$\begin{aligned} &\frac{1}{2}\partial_t\|\partial_y 2\Omega\gamma\psi\|^2 + \|\nabla\partial_y 2\Omega\gamma\psi\|^2 - \\ &-((2\Omega\gamma(Re - 2\Omega)(-\partial_y^2\varphi), \partial_y 2\Omega\gamma\psi) + \\ &+(-\boldsymbol{\varepsilon}\cdot 2\Omega\gamma(\mathbf{u}\cdot\nabla\mathbf{u}), 2\Omega\gamma\psi) = 0, \end{aligned}$$

$$\begin{aligned} -2\Omega\gamma\boldsymbol{\varepsilon}\cdot(\mathbf{u}\cdot\nabla\mathbf{u}) &= -2\Omega\gamma\partial_y(\boldsymbol{\delta}\varphi\cdot\nabla\partial_y\psi) - \\ &\quad -2\Omega\gamma\partial_y(f_2\partial_y(\partial_y\psi)) - \\ &\quad -2\Omega\gamma\partial_y((-\partial_y^2\varphi)\partial_z f_1). \end{aligned}$$

Since div $\boldsymbol{\delta}\varphi = 0$, $\partial_y\psi = 0$ at $z = \pm\frac{1}{2}$ it follows that

$$\begin{aligned} &(-\boldsymbol{\varepsilon}\cdot 2\Omega\gamma(\mathbf{u}\cdot\nabla\mathbf{u}), 2\Omega\gamma\psi) = \\ &= \int_{-\pi/\beta}^{\pi/\beta}\int_{-\frac{1}{2}}^{\frac{1}{2}} 4\Omega^2\gamma^2 f_2\partial_y^2\psi\partial_y\psi\,dz\,dy + \\ &\quad + \int_{-\pi/\beta}^{\pi/\beta}\int_{-\frac{1}{2}}^{\frac{1}{2}} 4\Omega^2\gamma^2(-\partial_y^2\varphi)\partial_z f_1\partial_y\psi\,dz\,dy \\ &= 4\Omega^2\gamma^2\int_{-\pi/\beta}^{\pi/\beta}\int_{-\frac{1}{2}}^{\frac{1}{2}}(-\partial_y^2\varphi)\partial_z f_1\partial_y\psi\,dz\,dy. \end{aligned}$$

Taking the scalar product over the layer of the f_1-row with $4\Omega^2\gamma^2 f_1$ we find

$$4\Omega^2\gamma^2 \int_{-\pi/\beta}^{\pi/\beta}\int_{-\frac{1}{2}}^{\frac{1}{2}} (\tilde{\mathbf{u}}\cdot\nabla\tilde{\mathbf{u}})_x f_1\,dz\,dy =$$

$$= 4\Omega^2\gamma^2 \int_{-\pi/\beta}^{\pi/\beta}\int_{-\frac{1}{2}}^{\frac{1}{2}} (\partial_{yz}\varphi\partial_y^2\psi + (-\partial_y^2\varphi)\partial_{yz}\psi) f_1\,dz\,dy,$$

$$= -4\Omega^2\gamma^2 \int_{-\pi/\beta}^{\pi/\beta}\int_{-\frac{1}{2}}^{\frac{1}{2}} \partial_y\varphi\partial_{yzy}\psi f_1\,dz\,dy$$

$$-4\Omega^2\gamma^2 \int_{-\pi/\beta}^{\pi/\beta}\int_{-\frac{1}{2}}^{\frac{1}{2}} \partial_y\varphi\partial_y^2\psi\partial_z f_1\,dz\,dy +$$

$$+4\Omega^2\gamma^2 \int_{-\pi/\beta}^{\pi/\beta}\int_{-\frac{1}{2}}^{\frac{1}{2}} (-\partial_y^2\varphi)\partial_{yz}\psi f_1\,dz\,dy,$$

$$= 4\Omega^2\gamma^2 \int_{-\pi/\beta}^{\pi/\beta}\int_{-\frac{1}{2}}^{\frac{1}{2}} \partial_y^2\varphi\partial_{yz}\psi f_1\,dz\,dy$$

$$-4\Omega^2\gamma^2 \int_{-\pi/\beta}^{\pi/\beta}\int_{-\frac{1}{2}}^{\frac{1}{2}} \partial_y\varphi\partial_y^2\psi\partial_z f_1\,dz\,dy +$$

$$+4\Omega^2\gamma^2 \int_{-\pi/\beta}^{\pi/\beta}\int_{-\frac{1}{2}}^{\frac{1}{2}} (-\partial_y^2\varphi)\partial_{yz}\psi f_1\,dz\,dy,$$

$$= 4\Omega^2\gamma^2 \int_{-\pi/\beta}^{\pi/\beta}\int_{-\frac{1}{2}}^{\frac{1}{2}} \partial_y^2\varphi\partial_y\psi\partial_z f_1\,dz\,dy.$$

Now we add the results we have got for the ψ-row and the f_1-row. There arises

$$\text{(II.13)}\qquad \left\{\begin{array}{c} \frac{1}{2}\partial_t(\|2\Omega\gamma\partial_y\psi\|^2 + \|2\Omega\gamma f_1\|^2)+ \\ +\|\nabla 2\Omega\gamma\partial_y\psi\|^2 + \|2\Omega\gamma\partial_z f_1\|^2- \\ -((2\Omega\gamma(Re-2\Omega)(-\partial_y^2\varphi), 2\Omega\gamma\partial_y\psi) = 0. \end{array}\right.$$

We form the scalar product over the layer of the first row in (II.1) with φ, then that one of the last row with f_2. This yields the expressions

$$\frac{1}{2}\partial_t\|\nabla\partial_y\varphi\|^2 + \|(-\Delta)\partial_y\varphi\|^2 - \frac{1}{\gamma}((-\partial_y^2)2\Omega\gamma\partial_y\psi,\varphi) + \\ +(\boldsymbol{\delta}\cdot(\mathbf{u}\cdot\nabla\mathbf{u}),\varphi) = 0,$$

$$\boldsymbol{\delta}\cdot(\mathbf{u}\cdot\nabla\mathbf{u}) = \boldsymbol{\delta}\cdot((\boldsymbol{\delta}\varphi+\mathbf{f})\cdot\nabla\boldsymbol{\delta}\varphi) + \partial_{yz}((-\partial_y^2\varphi)\partial_z f_2),$$

$$\int_{-\pi/\beta}^{\pi/\beta}\int_{-\frac{1}{2}}^{\frac{1}{2}}(\tilde{\mathbf{u}}\cdot\nabla\tilde{\mathbf{u}})_y f_2\,dz\,dy =$$

$$= \int_{-\pi/\beta}^{\pi/\beta}\int_{-\frac{1}{2}}^{\frac{1}{2}}((\partial_{yz}\varphi)\partial_y(\partial_{yz}\varphi)f_2 + (-\partial_y^2\varphi)\cdot \\ \cdot\partial_z(\partial_{yz}\varphi)f_2)\,dz\,dy,$$

$$= \int_{-\pi/\beta}^{\pi/\beta}\int_{-\frac{1}{2}}^{\frac{1}{2}}\partial_z\partial_y^2\varphi\partial_{yz}\varphi f_2\,dz\,dy -$$

$$- \int_{-\pi/\beta}^{\pi/\beta}\int_{-\frac{1}{2}}^{\frac{1}{2}}(-\partial_y^2\varphi)\partial_{yz}\varphi\partial_z f_2\,dz\,dy,$$

$$= -\int_{-\pi/\beta}^{\pi/\beta}\int_{-\frac{1}{2}}^{\frac{1}{2}}(-\partial_y^2\varphi)\partial_{yz}\varphi\partial_z f_2\,dz\,dy.$$

We add our results for the φ-row and the f_2-row. Then

$$\text{(II.14)}\qquad \begin{cases} \frac{1}{2}\partial_t(\|\nabla\partial_y\varphi\|^2 + \|f_2\|^2) + \|(-\Delta)\partial_y\varphi\|^2 + \\ +\|\partial_z f_2\|^2 - \frac{1}{\gamma}((-\partial_y^2)2\Omega\gamma\partial_y\psi,\varphi) = 0. \end{cases}$$

We set $\vartheta = 2\Omega\gamma\partial_y\psi$, $\tilde{f}_1 = 2\Omega\gamma f_1$, $\tilde{f}_2 = f_2$ and add (II.13, 14). We arrive at

$$\text{(II.15)}\qquad \begin{cases} \frac{1}{2}\partial_t(\|\nabla\partial_y\varphi\|^2 + \|\vartheta\|^2 + \|\tilde{f}_1\|^2 + \|\tilde{f}_2\|^2) \\ +(\|(-\Delta)\partial_y\varphi\|^2 + \|\nabla\vartheta\|^2 + \|\partial_z\tilde{f}_1\|^2 + \|\partial_z\tilde{f}_2\|^2)\cdot \\ \cdot\Big(1 - [\frac{1}{\gamma}((-\partial_y^2)\vartheta,\varphi) + \gamma 2\Omega(Re-2\Omega)\cdot((-\partial_y^2\varphi),\vartheta)]\cdot \\ \cdot[\|(-\Delta)\partial_y\varphi\|^2 + \|\nabla\vartheta\|^2 + \|\partial_z\tilde{f}_1\|^2 + \|\partial_z\tilde{f}_2\|^2]^{-1}\Big) = 0. \end{cases}$$

If $2\Omega(Re-2\Omega) < 0$ we set

$$\gamma = |2\Omega(Re-2\Omega)|^{-\frac{1}{2}}$$

and the terms within the first brackets [...] cancel. This proves assertion i) if $2\Omega(Re-2\Omega)<0$. As for $2\Omega(Re-2\Omega)>0$ we set

$$\gamma=(2\Omega(Re-2\Omega))^{-\frac{1}{2}}=R^{-\frac{1}{2}}.$$

(II.15) then apparently is nothing else but the equation which occurs when determining the energetic Rayleigh-number in the Bénard-problem under rigid boundaries (cf. [8, p.254]). This proves assertion i) if $2\Omega(Re-2\Omega)>0$. Concerning $2\Omega(Re-2\Omega)=0$ we set $\gamma=1$. As we know from [8, p.254])

$$\|(-\Delta)\partial_y\varphi\|^2+\|\nabla\vartheta\|^2\geq 2\sqrt{1707,...}\cdot|((-\partial_y^2)\vartheta,\varphi)|$$

with $\sqrt{1707,...}$ being the critical Rayleigh-number. This proves assertion ii). □

If $\Omega=0$ the functional F_2 neither contains ψ nor f_1. From the decay of the terms containing φ and f_2 one can conclude however that $\|\partial_y\psi(t)\|^2+\|f_1(t)\|^2$ also decays. This result is part of a more general assertion which says: If any plane parallel shear flow

$$\begin{pmatrix} f(z) \\ 0 \\ 0 \end{pmatrix},$$

whose profile f may thus be arbitrary, is perturbed by x-independent, y-periodic (socalled "transversal") disturbances, then it is unconditionally asymptotically stable against these, cf. [5, §24].

F_1 or F_2 for $\Omega\neq 0$ are equivalent to kinetic energy. If this one decays then the L^2-norms over the layer of higher order derivatives of the perturbation also decay, no matter how large the initial value is. This is a consequence of a more general 3D-theorem in [6].

Acknowledgement

One of the authors (R.K.) would like to thank B. Schmitt and F. Zimmermann for valuable comments on the material presented in this paper. He is also indebted to Deutsche Forschungsgemeinschaft (DFG) for financial support.

References

[1] Busse, F. H.: ber notwendige und hinreichende Kriterien fr die Stabilitt von Strmungen. ZAMM 50, T 173 - T 174(1970).

[2] Busse, F. H.: Strukturbildung in dissipativen Systemen. Course held during winter-term 1990/1991 at the University of Bayreuth.

[3] Galdi, G. P. and Rionero, S.: Weighted Energy Methods in Fluid Dynamics and Elasticity. Lecture Notes in Mathematics 1134. Springer: 1985.

[4] Galdi, G. P. and Padula, M.: A New Approach to Energy Theory in the Stability of Fluid Motion. Arch. Rational Mech. Anal. 110, 187-286(1990).

[5] Joseph, D. D.: Stability of Fluid Motions, Vol. I. Springer Tracts in Natural Philosophy 28. Springer: 1976.

[6] Kagei, Y. and Wahl, W. von: Asymptotic Stability of Higher Order Norms in Terms of Asymptotic Energy Stability for Viscous Incompressible Fluid Flows Heated from Below. To appear in the Japan J. Ind. and Appl. Math.

[7] Kagei, Y. and Wahl, W. von: Stability of higher norms in terms of energy stability for the Boussinesq-equations. Remarks on the asymptotic behaviour of convection-roll type solutions. Diff. Int. Eqs. 7, 921-948(1994).

[8] Schmitt, B. J. and Wahl, W. von: Monotonicity and Boundedness in the Boussinesq-Equations. Eur. J. Mech. B/ Fluids 12, 245-270(1993).

[9] Wahl, W. von: Necessary and Sufficient Conditions for the Stability of Flows of Incompressible Viscous Fluids. Arch. Rational Mech. Anal. 126, 103-129(1994).

[10] Wahl, W. von: The Boussinesq-Equations in Terms of Poloidal and Toroidal Fields and the Mean Flow. Lecture Notes. Bayreuther Math. Schriften 40, 203-290(1992).

Ralf Kaiser and Wolf von Wahl
Department of Mathematics
University of Bayreuth
D-95440 Bayreuth
Germany